# Lecture Notes in Computer Scienc

*Commenced Publication in 1973*
Founding and Former Series Editors:
Gerhard Goos, Juris Hartmanis, and Jan van Leeuwen

Richi Nayak   Mohammed J. Zaki (Eds.)

# Knowledge Discovery from XML Documents

First International Workshop, KDXD 2006
Singapore, April 9, 2006
Proceedings

Springer

Volume Editors

Richi Nayak
Queensland University of Technology
Faculty of Information Technology, GP, School of Information Systems
GPO Box 2434, Brisbane, QLD 4001, Australia
E-mail: r.nayak@qut.edu.au

Mohammed J. Zaki
Rensselaer Polytechnic Institute, Computer Science Department
Troy, NY 12180-3590, USA
E-mail: zaki@cs.rpi.edu

Library of Congress Control Number: 2006922467

CR Subject Classification (1998): H.2.8, I.2, H.3, F.1

LNCS Sublibrary: SL 3 – Information Systems and Application, incl. Internet/Web and HCI

ISSN 0302-9743
ISBN-10 3-540-33180-8 Springer Berlin Heidelberg New York
ISBN-13 978-3-540-33180-3 Springer Berlin Heidelberg New York



Springer is a part of Springer Science+Business Media

springer.com


Printed in Germany

Typesetting: Camera-ready by author, data conversion by Scientific Publishing Services, Chennai, India
Printed on acid-free paper SPIN: 11730262 06/3142 5 4 3 2 1 0

# Preface

The KDXD 2006 (Knowledge Discovery from XML Documents) workshop is the first international workshop running this year in conjunction with the 10th Pacific-Asia Conference on Knowledge Discovery and Data Mining, PAKDD 2006. The workshop provided an important forum for the dissemination and exchange of new ideas and research related to XML data discovery and retrieval.

The eXtensible Markup Language (XML) has become a standard language for data representation and exchange. With the continuous growth in XML data sources, the ability to manage collections of XML documents and discover knowledge from them for decision support becomes increasingly important. Due to the inherent flexibility of XML, in both structure and semantics, inferring important knowledge from XML data is faced with new challenges as well as benefits. The objective of the workshop was to bring together researchers and practitioners to discuss all aspects of the emerging XML data management challenges. Thus, the topics of interest included, but were not limited to: XML data mining methods; XML data mining applications; XML data management emerging issues and challenges; XML in improving knowledge discovery process; and Benchmarks and mining performance using XML databases.

The workshop received 26 submissions. We would like to thank all those who submitted their work to the workshop under relatively pressuring time deadlines. We selected ten high-quality full papers for discussion and presentation in the workshop and for inclusion in the proceedings after being peer-reviewed by at least three members of the Program Committee. Accepted papers were grouped in three sessions and allocated equal presentation time slots. The first session was on XML data mining methods of classification, clustering and association. The second session focused on the XML data reasoning and querying methods and query optimization. The last session was on XML data applications of transportation and security. The workshop also included two invited talks from leading researchers in this area. We would sincerely like to thank Tok wang Ling and Stephane Bressan for presenting valuable talks in the workshop program.

Special thanks go to the Program Committee members, who shared their expertise and time to make KDXD 2006 a success. The final quality of selected papers reflects their efforts.

Finally, we would like to thank Queensland University of Technology for providing us with the resources and time and the Indian Institute of Technology, Roorkee India for providing us with the resources to undertake this task. Last but least, we would like to thank the organizers of PAKDD 2006 for hosting KDXD 2006. We trust that you will enjoy the papers in this volume.

January 2006

Richi Nayak
Mohammad Zaki

# Organization

KDXD 2006 was organized by the School of Information System, Queensland University of Technology, Brisbane, Australia, in cooperation with PAKDD 2006.

## Workshop Chairs

| | |
|---|---|
| Richi Nayak | Queensland University of Technology, Australia |
| Mohammad Zaki | Rensselaer Polytechnic Institute, USA |

## Program Committee

Hiroki Arimura (Japan)
Giovanna Guerrini (Italy)
Jung-Won Lee (Korea)
Xue Li (Australia)
Yuefeng Li (Australia)
Chengfei Liu (Australia)
Marco Mesiti(Italy)
Ankush Mittal (India)
Shi Nansi (Australia)
Siegfried Nijssen (Netherlands)
Maria Orlowska (Australia)
Seung-Soo Park (Korea)
Wenny Rahayu ( Australia)
Michael Schrefl (Austria)
David Tanier (Australia)
Takeaki Uno (Japan)
Yue Xu (Australia)

# Table of Contents

## Keynote Papers



## XML Data Mining Methods



## XML Data Reasoning and Querying Methods

# Opportunities for XML Data Mining in Modern Applications, or XML Data Mining: Where Is the Ore?

Stephane Bressan, Anthony Tung, and Yang Rui

Department of Computer Science, School of Computing,
National University of Singapore
steph@nus.edu.sg

**Abstract.** We attempt to identify the opportunities for XML data mining in modern applications. We will try and match requirements of modern application managing XML data with the capabilities of the existing XML mining tools and techniques.

R. Nayak and M.J. Zaki (Eds.): KDXD 2006, LNCS 3915, p. 1, 2006.

# Capturing Semantics in XML Documents

Tok Wang Ling

Department of Computer Science, School of Computing,
National University of Singapore
lingtw@comp.nus.edu.sg

**Abstract.** Traditional semantic data models, such as the Entity Relationship (ER) data model, are used to represent real world semantics that are crucial for the effective management of structured data. The semantics that can be expressed in the ER data model include the representation of entity types together with their identifiers and attributes, n-ary relationship types together with their participating entity types and attributes, and functional dependencies among the participating entity types of relationship types and their attributes, etc.

Today, semistructured data has become more prevalent on the Web, and XML has become the de facto standard for semi-structured data. A DTD and an XML Schema of an XML document only reflect the hierarchical structure of the semistructured data stored in the XML document. The hierarchical structures of XML documents are captured by the relationships between an element and its attributes, and between an element and its subelements. Element-attribute relationships do not have clear semantics, and the relationships between elements and their subelements are binary. The semantics of n-ary relationships with $n > 2$ cannot be represented or captured correctly and precisely in DTD and XML Schema. Many of the crucial semantics captured by the ER model for structured data are not captured by either DTD or XML Schema. We present the problems encountered in order to correctly and efficiently store, query, and transform (view) XML documents without knowing these important semantics. We solve these problems by using a semantic-rich data model called the ***O***bject, ***R***elationship, ***A***ttribute data model for ***S***emi***S***tructured Data (ORA-SS). We briefly describe how to mine such important semantics from given XML documents.

R. Nayak and M.J. Zaki (Eds.): KDXD 2006, LNCS 3915, p. 2, 2006.

# Mining Changes from Versions of Dynamic XML Documents

Laura Irina Rusu[1], Wenny Rahayu[2], and David Taniar[3]

[1,2] LaTrobe University, Department of Computer Science & Computer Eng, Australia
lirusu@students.latrobe.edu.au
wenny@cs.latrobe.edu.au
[3] Monash University, School of Business Systems, Clayton, VIC 3800, Australia
David.Taniar@infotech.monash.edu.au

**Abstract.** The ability to store information contained in XML documents for future reference becomes a very important issue these days, as the number of applications which use and exchange data in XML format is growing continuously. Moreover, the contents of XML documents are dynamic and they change across time, so researchers are looking to efficient solutions to store the documents' versions and eventually extract interesting information out of them. This paper proposes a novel approach for mining association rules from changes between versions of dynamic XML documents, in a simple manner, by using the information contained in the consolidated delta. We argue that by applying our proposed algorithm, important information about the behaviour of the changed XML document in time could be extracted and then used to make predictions about its future performance.

## 1 Introduction

The increasing interest from various applications in storing and manipulating their data in XML format has determined, during the last few years, a growing amount of research work, in order to find the most effective and usable solutions in this respect. One main focus area was XML warehousing [9, 10], but a large volume of work have been also concentrating on the issue of mining XML documents [7, 8, 11]. The later one evolved in a quite sensitive issue, because the users became interested not only in storing the XML documents in a very efficient way and accessing them at any point in time, but also in getting the most of the interesting information behind the data.

In addressing first part of the problem, i.e. XML warehousing, we have identified at least two types of documents which could be included in a presumptive XML data warehouse: *static XML documents*, which do not change their contents and structures in time (e.g. an XML document containing the papers published in a proceedings book) and *dynamic XML documents*, which change their structures or contents based on certain business processes (e.g. the content of an on-line store might change hourly, daily or weekly, depending on the customer behavior). While the first category of XML documents was the subject of intense research during the recent years, with various methods for storing and mining them being developed, there is still work to be done in finding efficient ways to store and mine dynamic XML documents [1].

R. Nayak and M.J. Zaki (Eds.): KDXD 2006, LNCS 3915, pp. 3 – 12, 2006.

The work in this paper continues the proposal made in [1], visually grouped in the general framework presented in Figure 1. In this framework, we focused on both warehousing and mining dynamic XML documents, in three main steps, i.e. (*i*) storing multiple versions of dynamic XML documents (Fig. 1A), (*ii*) extracting historic changes for a certain period of time (Fig.1B) and (*iii*) mining the extracted changes (Fig.1C) to obtain interesting information (i.e. association rules) from them.

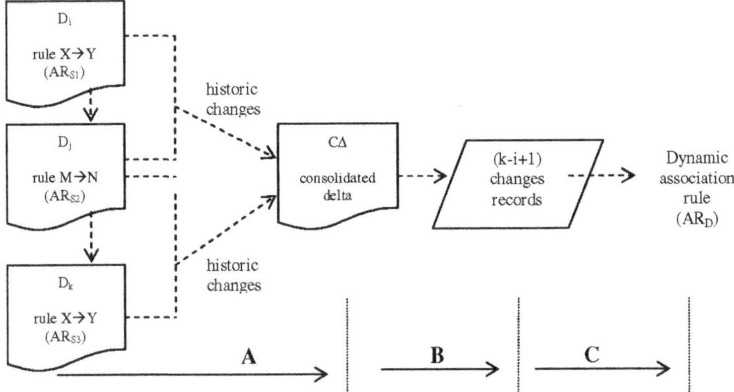


**Fig. 1.** A visual representation of the mining historic changes process, using consolidated delta

In this paper, we are focusing on the part C of the above mentioned framework, i.e. extracting association rules from changes affecting dynamic XML documents. We believe this knowledge would be very useful in determining if there are any relationships between changes affecting different parts of the documents and making predictions about the future behaviour of the document.

## 2 Related work

To our knowledge, there is no much work done in the area of mining changes between versions of dynamic XML documents. The existing work is more focused on determining interesting knowledge (e.g. frequently changing structures, discovering association rules or pattern-based dynamic structures) from the multiple versions of the document themselves, not from the actual changes happened in the specified interval of time. We detail below some of this work, noting in the same time that the list of examples is nor complete or exhaustive.

In [2], the authors focus on extracting the FCSs (Frequently Changing Structures). They propose an H-DOM model to represent and store the XML structural data, where the history of structural data is preserved and compressed. Based on the H-DOM model, they present two algorithms to discover the FCSs.

X-Diff algorithm is proposed in [3] and it deals with unordered trees, defined as trees where only the ancestor relationship is important, but not the order of the siblings. This approach is considered to be better and more efficient for the purpose of database applications of the XML. In [3], changes in a XML document over the time are determined by calculating the *minimum-cost edit script*, which is a specific sequence of operations which can transform the XML tree from the initial to the final phase, with the lowest possible cost. In introduces the notion of *node signature* and a new type of matching between two trees, corresponding to the versions of a document, utilized to find the minimum cost matching and cost edit script, able to transform one tree into another.

Another algorithm, proposed by [4], deals with the unordered tree as well, but it goes further and does not distinguish between elements and attributes, both of them being mapped to a set of labeled nodes.

In [5], the authors focus on discovering the pattern-based dynamic structures from versions of unordered XML documents. They present the definitions of *dynamic metrics* and *pattern-based dynamic structure mining* from versions of XML documents. They focus especially on two types of pattern-based dynamic structures, i.e. *increasing dynamic structure* and *decreasing dynamic structure*, which are defined with respect to dynamic metrics and used to build the pattern-based dynamic structures mining algorithm.

## 3 Problem Specification

To exemplify the problem, in Figure 2 we present one XML document, at the time $T_0$ (the initial document), followed by three versions, at three consecutive moments of time, i.e. $T_1$, $T_2$ and $T_3$. Each version brings some changes to the previous one, visually represented by the dotted lines.

One technique for storing the changes of a dynamic XML document (i.e. which changes its context in time) was proposed in [1]. Three main features of this technique are: (*i*) the resulting XML document is much smaller in size than the sum of all versions' sizes; (*ii*) it allows running a simple algorithm to extract any historic version of the document and (*iii*) the degree of redundancy of the stored data is very small, only the necessary information for quick versioning being included.

By running the consolidated delta algorithm [1], we obtain a single XML document containing the historic changes on top of the initial document. We resume here the main steps in building the consolidated delta and few important concepts, for a document changing from the version $D_i$ (at time $T_i$) to version $D_j$ (at time $D_j$):

- unique identifiers are assigned for the new inserted elements in the $D_i$ version;
- version $D_j$ is compared with the previous one $D_i$ and for each changed element in the $D_j$ version, a new child element is inserted in the consolidated delta, namely <stamp>, with two attributes: (*a*) "time" , which contain the $T_i$ value ( e.g. month, year etc) and (*b*) "delta" , which contain one of *modified, added, deleted* or *unchanged* values, depending on the change detected at the time $T_i$; there are some rules to be observed when adding the <stamp> elements [1];
- the $D_i$ version is removed from the data warehouse, as it can be anytime recreated using the consolidated delta. The $D_j$ version is kept until a new version arrives or until a decision to stop the versioning process is taken; *D*j will be removed after the last run of the consolidated delta algorithm;
- at the end of the process, the consolidated delta will contain enough historical information to allow for versioning.

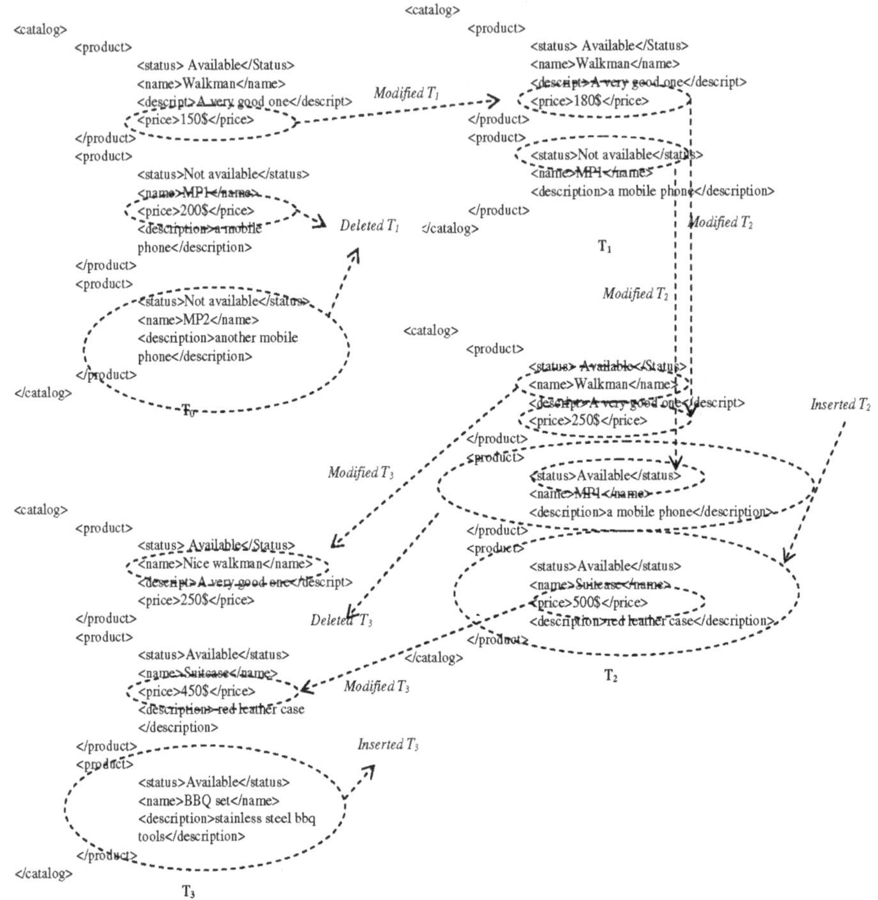


**Fig. 2.** The "catalog.xml" document in four consecutive versions

We need to mention that the $D_0$ version of the XML document (i.e. the initial one) will be included in the initial consolidated delta; from that point, only the changes for any subsequent version will be recorded in the consolidated delta as described above.

After running the consolidated delta algorithm [1] to capture all the changes affecting the running example document in period $T_0 - T_3$, we will obtain an XML document where each initial element from $D_0$ has attached a history of its changes. Note that, if an element was either deleted at updated at a time $Ti$, $0<i<3$, its children do not have attached any stamp elements for that specific time and this helps in limiting as much as possible the degree of redundancy of the data stored in the consolidated delta.

In our working example, the changes affecting the initial XML document during the consecutive transformations from a version to another are presented in Table 1.

**Table 1.** The list of changes for the working example, for three consecutive versions of an XML document

| | |
|---|---|
| $T_0 \rightarrow T_1$ | Price – modified; Product – deleted; Price – deleted |
| $T_1 \rightarrow T_2$ | Product – inserted; Price – modified; Status – modified |
| $T_2 \rightarrow T_3$ | Name – modified; Product – deleted; Price – modified; Product – inserted |

If, in our working example, we consider the sets of changes in periods $T_0 \rightarrow T_1$, $T_1 \rightarrow T_2$ and $T_2 \rightarrow T_3$ (see Table 1) as transactions, it can be noticed that the pairs "Price-modified" and "Product-inserted" appear in 2 of the 3 (66%) of the transactions. If the minimum support required is set at a level lower than 66%, the association rule extracted would be: "when a *price* element is modified, a *product* is inserted as well, and this happens in 66% of the sets of changes appearing from a version to another".

This paper proposes to build a generic algorithm for extracting association rules from the changes which affect dynamic XML documents, i.e. to discover if there are any relationships between modifications, deletions or insertions of some elements or another. As exemplified above, the resulting rules could be very informative about how parts of the document are changing together so the user can make predictions about the future behaviour of the dynamic XML document.

## 4 Mining Changes – The Proposed Algorithm

The algorithm for mining changes from historic versions of dynamic XML documents is an improved Apriori one, redesigned to be applicable to XML documents; it has a preparation step and four main working steps, as follows:

```
For each Ti, 0<i<n
 Get all nodes with timestamp Ti
  For each node with timestamp Ti and delta not "unchanged"
   If the delta is "modified" or "inserted"
    If the node has no other children elements except stamp
      Record timestamp, value and delta
    End if
   Else  ' i.e. delta is "deleted"
    Record timestamp and delta
   End if
  Next
Next
```

**Fig. 3.** Generic algorithm for extracting historic changes (ECD) from consolidated delta document

**Preparation step:**
Because we want to mine the actual changes which influenced the initial XML document, we use the consolidated delta to extract the set of changes, using the algorithm proposed in Figure 3. We will name the resulting document ECD (the extracted changes document) further in the paper.

For each moment of time *Ti* when the document was changed, $0<i<=n$, we will have a *transaction* containing the elements changed at the time *Ti*. Each combination {element – change} will actually become an *item* in our mining algorithm.

During this step, also count the number of transactions in the ECD. It will be relevant for calculating the support of the association rules discovered.

**Step 1:**
During this step, a new XML document is built, to store the number of modifications, deletions, insertions for each element $E_i$ in ECD, together with the associated support for each pair "element-change'. The document will be named MCdoc (matrix of changes in the document), further in the paper.

The number of modifications, insertions and deletions will be stored as attributes of each element in MCdoc. Each time a new element is found in the extracted changes document, the *modified, inserted* or *deleted* attribute will take value 1. If the same element is found again to be changed during a different transaction, the corresponding attribute will be updated to reflect the current number of changes; the support of the pair {element – change} will be updated too, with regard to the total number of transactions (see preparation step).

If during the preparation step, we also include the paths of the elements in the extracted changes document, the algorithm will be able to identify only the distinct elements and their changes, so elements with same name but different positions in the hierarchy will be recorded separately. The algorithm for Step 1 is detailed in Figure 4.

In our working example, the total number of changes extracted is 10 and the result of applying this step is the document in Figure 6.

```
For each element Ei in ECD
    Search MCdoc for the element Ei
    If not found
        Include it in the MCdoc, with value 1 to the appropriate argument
    Else
        'check if the path is different
        If path attribute is different
            Include it in the MCdoc, with value 1 to the appropriate argument
        Else
            Update the appropriate argument to reflect the change attached
            Update the support of the pair "element – change'
        End if
    End if
Next element Ei
```

**Fig. 4.** Step 1 of the proposed algorithm, i.e. building the document with changing items and their support from the ECD (extracted changes document)

```
Set L = set of large 1-itemsets
For each element in MCDoc
  If sup_modif > min_sup
        Add {"element" – modif} pair to L
  If sup_insert > min_sup
        Add {"element" – inserted} pair to L
  If sup_delete > min_sup
        Add {"element" – deleted} pair to L
Next element
```

**Fig. 5.** Step 2 of the proposed algorithm, i.e. determining the large 1-itemsets

**Step 2:**
The 1-large itemsets will be extracted from the document build at step 1, i.e. those items (element - change pairs) which have the support in ECD higher than the min_sup set at the beginning of the process (see Figure 5 for the algorithm).

In our example, we set the minimum support to 20%, so the frequent 1-itemsets are three: {price - modified}, {product - inserted} and {product-deleted} (see dotted lines in Figure 6). We emphasise again the fact that, in our proposed algorithm, not the actual elements from the initial document are the items in transactions, but the combinations (pairs) element – change.

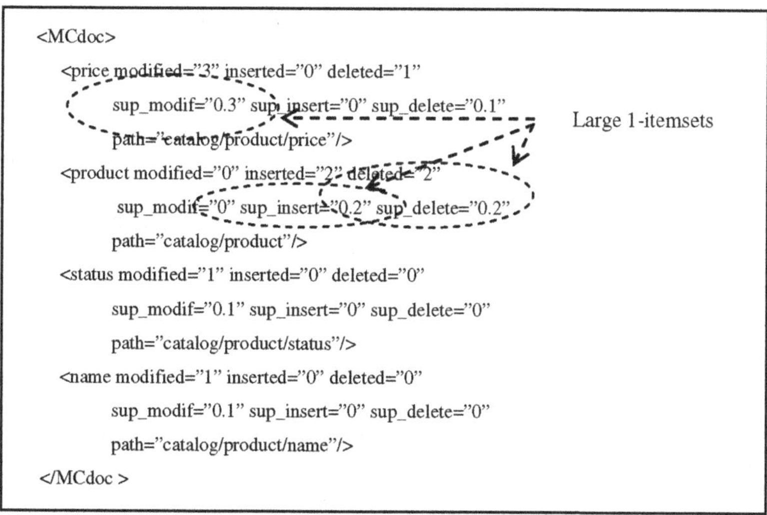


**Fig. 6.** The document resulted by applying Step 1 of the proposed mining changes algorithm, highlighting the large 1-itemsets in the working example

**Step 3:**
Similar with the Apriori-based algorithm, the *k*-itemsets (*k*>1) are built starting from the 1-itemsets; for each of them the support is calculated with respect to the total

number of changes from ECD. This step is repeated until all the large *n*-itemsets are found. This step will be influenced by the observation that any larke *k*-itemset (i.e. which has a support greater than minimum required) needs to have all its subsets large. In figure 7, we show a general *k* step.

In our example, the 2-itemset {price – modified, product - inserted} has the support=0.66, because both pairs appear in two out of three transactions extracted.

```
For each large (k-1) itemset in L_k-1
  Extend to k itemset by adding a 1-large itemset
    'calculate its support in ECD (extracted changes document)
    For each transaction in ECD
      Set bPairFound=false
      For each pair {element-change} in the k itemset
          If (transaction//element and transaction//change) not null
              Set bPairFound= true
          Else
              Set bPairFound= true
      Next pair
      If bPairFound=true 'found all pairs from k itemset
          Update S_k - support of k itemset
    Next transaction
    If S_k > min_sup then add k itemset to L_k - the list of large k itemsets
Next (k-1) itemset
```

**Fig. 7.** Step 3 of the proposed algorithm, i.e. determining the large k-itemsets from the (k-1) large itemsets

**Step 4:**
Based on the large *n*-itemsets extracted at Step 3, we determine the association rule and calculate their confidence. In the above working example, the rule extracted is "if a price element is modified, a product is inserted as well, and this happens in 66% of the sets of changes appearing from a version to another" and it will have a confidence of 66%.

## 5 Evaluation of Experimental Results

To evaluate the proposed approach we used the consolidated delta obtained from successive versions of the same XML documents of various sizes, i.e. 25kb, 63kB, and 127kB, data being downloaded from the SIGMOD dataset [6]. Firstly, we built a number of versions for each type of document by using a changes simulator, created by us, which takes as input the $D_i$ version of the document ($0<i<n$) and returns a modified $D_{i+1}$ version, where the desired percentages of deletions, insertions and modifications can be controlled through a user-friendly interface and elements to be changed are randomly chosen. For each new version of each document, we calculate the corresponding consolidated delta, observing the rules in [1];

By applying the algorithm proposed in this paper and implemented in Visual Basic, we got a number of association rules and we made measurements of the running time, for different values of the minimum support required. The graph in Figure 8 shows the time of running for each consolidated delta obtained after few sets of changes (number showed in parenthesis on the graph legend) and for different values of minimum support required.

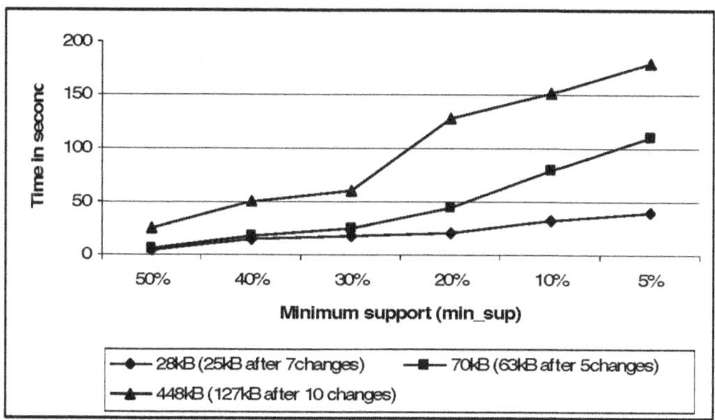


**Fig. 8.** Running time for three different sized consolidated delta and six different percentages of minimum support imposed

As it can be noticed from the graph, the smaller consolidated delta has the best results as time of running, as one would expect and even for large XML documents (as it is the 448kB consolidated delta) the time is kept under 3 minutes. We need to mention that the first two steps, i.e. preparation step (when the ECD - extracted changes document is built) and step 1 (when the large 1-itemsets are identified) are the most expensive ones, as time and processor resources. Our future work is to explore the possibilities of making this steps more efficient, so the overall performance of the algorithm to be improved.

## 6 Conclusions

In conclusion, this paper presents a novel approach for mining changes extracted from versions of dynamic XML documents, by looking into the actual changes and into the associations between them. The motivation for this research was that the user not only needs to know which are, for example, the most changing parts of the document, but also which are the relationships between the changes of the document's parts, e.g. modifications of some parts of the document might be related with insertions of some new parts or with deletions of other parts. The information extracted would be very useful to predict the future behaviour of a dynamic XML document. We hence propose an algorithm to mine these changes, in few clear steps, with examples easy to understand and replicate.

## References

1. Rusu, L.I., Rahayu W., Taniar D., "Maintaining Versions of Dynamic XML Documents", *Proceed. of The 6th International Conference on Web Information Systems Engineering (WISE 2005)*, New York, LNCS 3806, pp. 536-543, 2005
2. Zhao, Q., Bhowmick, S.S., Mohania, M., Kambayashi, Y., "FCS Mining: Discovering Frequently Changing Structures from Historical Structural Deltas of Unordered XML", *In Proceedings of the 13th Conference on Information and Knowledge Management (CIKM 2004)*, pp. 188-197
3. Wang Y., DeWitt D.J., Cai J.Y., "X-Diff: An Effective Change Detection Algorithms for XML Documents", *In Proceedings of ICDE 2003*, pp.519-530, IEEE Computer Society, 2003
4. Zhao, Q., Bhowmick, S.S., Mohania, M., Kambayashi, Y., "Discovering Frequently Changing Structures from Historical Structural Deltas of Unordered XML", *Proceedings of ACM CIKM'04*, pp.188-197, November 8-13, Washington, US, 2004
5. Zhao, Q., Bhowmick, S.S., Mandria, S., "Discovering Pattern-based Dynamic Structure from Versions of Unordered XML Documents", In *Proceedings of the 6th International Conference on Data Warehousing and Knowledge Discovery (DaWak 2004)*, pp.77-86, Zaragoza, Spain, September 1-3, 2004
6. www.cs.washington.edu/datasets - SIGMOD XML dataset
7. Yin, M., Goh, D.H-L, Lim, E-P., and Sun, A., "Discovery of Content Entities from Web Sites Using Web Unit Mining", *International Journal of Web Information Systems*, 1(3):123-136, 2005.
8. Zhou, B., Hui, S.C., and Fong, A.C.M., "A Web Usage Lattice Based Mining Approach for Intelligent Web Perzonalization", *International Journal of Web Information Systems*, 1(3):137-146, 2005.
9. Quang, N.H., Rahayu, W., "XML Schema Design Approach", *International Journal of Web Information Systems*, 1(3):161-178, 2005.
10. Rusu L.I., Rahayu W., Taniar D., "A methodology for building XML data warehouses", *International Journal of Data Warehousing and Mining, vol.1, no.2*, pp.67-92, April - June 2005
11. Feng.L, Dillon T., "An XML-enabled data mining query language: XML-DMQL", *International Journal of Business Intelligence and Data Mining, vol 1, no 1*, 22-41, 2005-11-30

# XML Document Clustering by Independent Component Analysis

Tong Wang[1], Da-Xin Liu[1], and Xuan-Zuo Lin[2]

[1] Department of Computer Science and Technology,
Harbin Engineering University, China
Wangtong@hrbeu.edu.cn
[2] Northeast Agriculture University, Harbin, China
xuanzuolin@sina.com

**Abstract.** When XML documents are clustered, the high dimensionality problem will occur. Independent Component Analysis (ICA) can reduce dimensionality and in the meanwhile find the underlying latent variables of XML structures to improve the quality of the clustering. This paper proposes a novel strategy to cluster XML documents based on ICA. According to *D_path* extracted from XML trees, the document was at first represented as Vector Space Model (VSM).Then ICA is applied to reduce the dimensionality of document vectors. Furthermore, document vectors are clustered on this reduced Euclidean Space spanned by the independent components. The experiments show that ICA can enhance the accuracy of the clustering with stable performance.

## 1 Introduction

XML is becoming the standard web data exchange format. Many research efforts are currently devoted to support the storage and retrieval of large collections of such documents. Our research is driven by the hypothesis that closely associated documents tend to be relevant to the same requests, thus grouping similar documents can accelerate the search [1]. Many researchers [2][3] measure structural similarity using the "edit distance" between tree structures. However, the edit distance between two documents has time complexity at least $O(n^2)$ and the algorithm requires computing the distance for each document-pair. Thus, it is unsuitable for a collection of large documents. In this paper, we represent XML documents using path sequences. The representation can embody the structural information with lower complexity.

Moreover, it is common to extract thousands of different words or features from text document in order to represent the vectors. The high dimensionality of natural text is often referred to as the "curse of dimensionality". In the context of clustering, the commonly used distance measure between data points will lose discriminative power gradually as the number of dimensions increases for the given dataset. It has been shown that, in a high dimensional space, data points almost always have equal distance to each other for various data distributions and distance functions [4]. To solve the high-dimensionality problem, various reduction dimension methods [5, 6] have been applied for clustering.

R. Nayak and M.J. Zaki (Eds.): KDXD 2006, LNCS 3915, pp. 13 – 21, 2006.

Independent Component Analysis (ICA) has gained widespread attention in signal processing, face recognition [7], etc. However, there are only few works in which ICA is applied to text applications such as topic discovery in temporal text [11], and unsupervised identification of linguistic features such as parts of speech [9]. In order to mine intrinsic structure of documents in higher-order statistics, this paper applied ICA to reduce dimension of vector space. To the best of author's knowledge, this paper is the first to introduce ICA into XML document clustering.

In this paper, we propose a novel clustering strategy for **XML** documents using **ICA** (ICAXC). Based on $D\_path$, we at first represent documents as vectors in VSM. Furthermore, we get the independent components of document matrix and cluster vectors in the reduced space spanned by ICs. The remainder of the paper is organized as follows: section 2 is the feature extraction; section 3 introduced the ICAXC technique; we analyze experiment results in section 4 and conclude in section 5.

## 2 Vector Representation

XML document can be viewed as a labeled tree. In our case, we define here XML document tree $d$ .

**Definition 1. XML document tree:** Suppose a countable infinite set $E$ of element labels (tags), a countable infinite set $A$ of attribute names. An *XML document tree* is defined to be $d = (V, lab, ele, att, v_r)$ where $V$ is a finite set of nodes in *d*; *lab* is a function from $V$ to $E \cup A$; *ele* is a partial function from $V$ to a sequence of $V$ nodes such that for any $v \in V$, if *ele(v)* is defined then $lab(v) \in E$; *att* is a partial function from $V \times A$ to $V$ such that for any $v \in V$ and $l \in A$, if $att(v,l) = v_1$, then $lab(v) \in E$ and $lab(v_1) = l$; $v_r$ is a distinguished node in $V$ called root of *d*, $lab(v_r) = root$.

Figure 1 shows an example of XML document trees. The model is a rooted, directed, and unordered tree. A path in $d$ is a sequence of nodes $v_1, v_2, v_3, ..., v_n$, through which we can traverse step by step in $d$. In addition, there exists one and only one path from node $v_i$ to node $v_j$ for each $v_i$ and $v_j$, provided $v_i \neq v_j$.

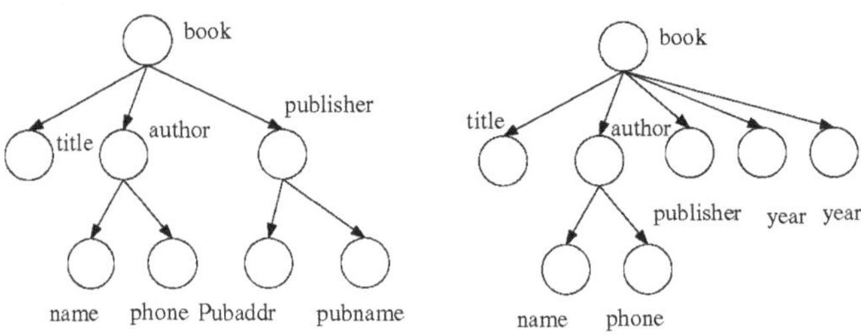


**Fig. 1.** XML document tree $d_1$ (left) and $d_2$ (right)

**Definition 2. Path Sequence:** Consider an XML document tree, $d$. The path sequence of $v_i$ is an ordered sequence of tag names from $v_r$ to $v_i$, written $D\_path^d_{v_i}$: $D\_path^d_{v_i} = \{v_0, v_1, ..., v_q\}$, where $v_k \in V$ ,k$\in$[1...q].

Given node $v_i$ and $v_p$, we define $v_i$ is *nested* in $v_p$ w.r.t. $i < p \wedge v_{i,} \quad v_p \in D\_path^d_{v_i}$. Note that $D\_path$ describes not only the XML structure but also hierarchical information of $d_k$, for it shows how $v_i$ is nested in $d_k$. Since an XML document $d_k$ consists of many $D\_path^{d_k}_{v_i}$ sequences, it can be expressed as follows:

$$d_k = \{D\_path^{d_k}_{v1}, D\_path^{d_k}_{v2}, ..., D\_path^{d_k}_{vw}\} \ . \tag{1}$$

We employ the Vector Space Model (VSM)[10] and consider a document collection, matrix $D_{m\times n} = (d_1, d_2 ... d_n)$, where n is the cardinality of the collection and m is the number of $D\_paths$ extracted from D. Each column is an m-dimensional vector relevant to document $d_k$, k$\in$[1...n] and every row corresponds to one $D\_path$. The $i_{th}$ row of $d_k$ is the number of the corresponding $D\_path$ occurrences if $D\_path$ exists in $d_k$; otherwise, the $i_{th}$ row of vector is 0. This arrangement is analogous to standard term-document matrix used in latent Semantic Indexing (LSI) literature [12].

**Table 1.** Representation of XML document D=( $d_1$, $d_2$ ), where $d_1$ and $d_2$ are given in figure 1

| *D_Path* | Real path sequence | $d_1$ | $d_2$ |
|---|---|---|---|
| $D\_path_1$ | <{book},{title}> | 1 | 1 |
| $D\_path_2$ | <{book},{author},{name}> | 1 | 1 |
| $D\_path_3$ | <{book},{author},{phone}> | 1 | 1 |
| $D\_path_4$ | <{book},{publisher},{pubaddr}> | 1 | 0 |
| $D\_path_5$ | <{book},{publisher}> | 0 | 1 |
| $D\_path_6$ | <{book},{year}> | 0 | 2 |
| $D\_path_7$ | <{book},{publisher},{pubname}> | 1 | 0 |

Let us take XML document collection D= ( $d_1$, $d_2$ ) as an example. As is shown in table1, $d_1$ $\{1,1,1,1,0,0,1\}^T$ and $d_2$ $\{1,1,1,0,1,2,0\}^T$. In the context of clustering, we choose to compute the Distance metric via cosine Distance [17].

$$Dis\tan ce(d_i, d_j) = \frac{(d_i)^T d_j}{\| d_i \| \cdot \| d_j \|} \ . \tag{2}$$

# 3 Document Clustering Based on ICA

The proposed **XML** Clustering strategy based on **ICA** (called **ICAXC**) consists of 3 stages: vector representation, feature transformation with ICA and clustering with C-means method. Section 2 has introduced the feature extraction and vector representation. In this section, we will show how ICA reduces the dimensionality of vector space and finds the latent variables. Then, standard C-means method is employed to cluster in reduced Euclidean space.

## 3.1 Background

Independent Component Analysis (ICA) is a statistical and computational technique for revealing hidden factors that underlie sets of random variables or signals. It is a general-purpose statistical technique, which tries to linearly transform the original data into components that are maximally independent from each other in a statistical sense. ICA has enjoyed good success in many different areas:

Consider, for example, electrical recordings of brain activity as given by an electroencephalogram (EEG). The EEG data consists of recordings of electrical potentials in many different locations on the scalp. These potentials are presumably generated by mixing some underlying components of brain activity. This situation is quite similar to the cocktail-party problem: we would like to find the original components of brain activity, but we can only observe mixtures of the components. ICA can reveal interesting information on brain activity by giving access to its independent components.

A very different application of ICA is on feature extraction. A fundamental problem in digital signal processing is to find suitable representations for image, audio or other kind of data for tasks like compression and denoising. Data representations are often based on (discrete) linear transformations. Standard linear transformations widely used in image processing are the Fourier, cosine transforms etc. Each of them has its own favorable properties [13].

However, until very recently, there are only a few experimental works in which ICA is applied to text applications [11][14]. ICA has been compared to LSI in producing representations better aligned with the grouping structure of the given text [14]. An extension of standard ICA to streaming data has been used successfully for identifying topics in a dynamical textual environment, i.e., chat room conversation streams [11]. So far, the other applications of ICA to text are still few.

## 3.2 Dimensionality Reduction of Document Matrix

When XML documents are clustered, the tags from the XML provide the structural and semantic information, which can improve the discriminative ability of each XML documents. Since $D\_path$ is nested in XML structures, there is close correlativeness and redundancy between these paths. Thus, ICA can help for mining the hidden relevancy from the deeper statistical level. Actually, the process of dimension reduction via ICA is actually the process of feature transformation in document collection, matrix $D_{m \times n}$. The lower dimensional space is often believed to represent the underlying latent structure or features in the matrix. Such a transformation can either guarantee a good degree of distance preservation among vectors or generate statistically more independent components of the original dataset.

Independent component analysis was originally developed to deal with problems that are closely related to the cocktail-party problem. However, in this paper, ICA procedure is described for document clustering.

We assume each observed data (in our case, a document $d_i$) being generated by a mixing process of statistically independent components (latent variables $s_i$). According to Vector Space Model, the noise-free mixing model can be written as

$$D_{m\times n} = A_{m\times k} \cdot S_{k\times n} . \tag{3}$$

where A is referred to as the mixing matrix. Suppose the inverse of matrix A is immixing matrix $W$, the independent components can be expressed as

$$S_{k\times n} = W_{k\times m} \cdot D_{m\times n} . \tag{4}$$

where $W$ is the projection matrix that projects $D$ from $m$ dimensional space to a lower $k$ dimensional space( $k \le m$ ).

In our application, the task of ICA is to use documents matrix $D = (d_1, d_2, ..., d_n)$ to estimate the mixing matrix $A$ and the independent components, $S$ , which represents the new documents collection. Afterwards, clustering is operated on this space spanned by the independent components.

The most commonly used implementation is fastICA [13], which is known to be robust and efficient in detecting the underlying independent components in the data for a wide range of underlying distributions [8]. The mathematical details of fastICA can be found in [11], which is not discussed here. FastICA has two pre-processing steps: centering and whitening. In our experiment, the most time consuming part of fastICA is the whitening, which can be computed with *SVDs* in Matlab™. Based on principal components of the matrix $D$ obtained in the whitening, fastICA algorithm then iterates to find one independent component each time by Negentropy-maximization [13].

### 3.3 Clustering Method

After dimension reduction, we discuss the use of the clustering algorithm, which is employed to cluster vectors in the reduced Euclidean space. We choose to use C-means, since C-means is the most popular clustering algorithms used in text clustering [15] and its efficiency, with time complexity $O(ntk)$, where n is the size of dataset, k is the clusters and t is the circle time. Besides, recent studies have shown that partitional clustering algorithms are more suitable for clustering large datasets than other clustering algorithms [19].

C-means algorithm is a simple, partitional clustering algorithm based on the firm foundation of analysis of variances. It clusters a group of data vectors into a predefined number of clusters. It starts with randomly initial cluster centroids and keeps reassigning the data objects in the dataset to cluster centroids based on the similarity between the data object and the cluster centroid. The reassignment procedure will not stop until a convergence criterion is metï(e.g., the fixed iteration number, or the cluster result does not change after a certain number of iterations).

# 4 Experimental Results and Analysis

We conducted the experiments on a workstation of 1.5GHz Intel Pentium 4 machine with 512 MB main memory.

## 4.1 Dataset

We choose a variety of XML datasets including two widely used real datasets and one synthetic dataset, Xmark. One real dataset is obtained from DBLP [16], the bibliographical data of scientific conferences and journals; the other is Swiss Prot, a real-life data set with annotations on proteins; Xmark, a synthetic dataset that models transactions on an on-line auction site. Compared with DBLP, the data in Xmark is relatively tilted and sparse, with more complex structures.

The test subset of DBLP we used consists of 10 different ACM Journals. Each journal with 100 documents is grouped, denoted by $G_i, 1 \le i \le 10$. We mix these documents together and cluster them for our test. In the context of clustering, we can also produce 10 categories, denoted by $C_i, 1 \le i \le 10$. Similarly, the subset of Protein set contains 1324 document that have been classified into 54 categories.

For the synthetic dataset, Xmark, our experiment is based on the hypothesis that the documents with the same DTD will be clustered in the same class. When we generate files using Xmark, the scale parameter of Xmark is 0.2. That is, each generated document is 20M or so. We input 5 DTD (Data Type Definition) documents [18] and for each DTD generate 20, 40, 60, 80, 100 XML documents, respectively. The five generated datasets are denoted as Xmark1, Xmark2, Xmark3, Xmark4 and Xmark5, respectively.

## 4.2 Measurement of Clustering Accuracy

In order to measure the clustering accuracy, we take the DBLP as an example. As mentioned above, the groups we specify beforehand are denoted by $G_i$, and the final clustered groups in the experiments are denoted by $C_i$. The $\delta$ function is given by

$$\delta(d_1, d_2, C_i) = \begin{cases} 0, \ if \ \exists j, d_1, d_2 \in G_j \\ 1, \ if \ \neg\exists j, d_1, d_2 \in G_j \end{cases} . \tag{5}$$

where $d_1$ $d_2$ are documents from $C_i$ category. To quantify the clustering accuracy of the ICAXC technique, we define Classified Error Rate (CER) as follows.

$$CER = \frac{\sum_{i} \sum_{m,n \in C_i \wedge m \neq n} \delta(m,n,C_i)}{\sum_{i} [i \times (i-1)/2]}. \tag{6}$$

If there is no pair of documents occurring in both C and G classes, the error rate will reach the maximum value, e.g., combination $C_i^2 = i \times (i-1)/2$. CER is a relative error rate value, $0 \leq CER \leq 1$. A lower CER value would indicate that the hidden variable discovered by the clustering is more informative of, or more useful in recovering, the original classification.

### 4.3 Experimental Analysis

To test the performance of the proposed strategy, we also implement the naïve clustering method and represent document vector using *D_path* sequences. In the third stage of ICAXC, we choose the same clustering method C-means as naïve method does. Besides, the documents were parsed into labeled trees via the parser developed by Zhang et al [20] in pre-process.

All tests are under Matlab 6.5.1 environment. The matlab code for fastICA is obtained from [13]. The C-means procedures are taken directly from Matlab toolboxes. In the stage of standard C-means procedure, the choice of k is often ad hoc, larger than the number of classes in general. In our case, we choose the class number. Since C-means is sensitive to the input order of vectors, we did each experiment several times and obtained the mean of CER. Fig.2 shows the results of the two methods.

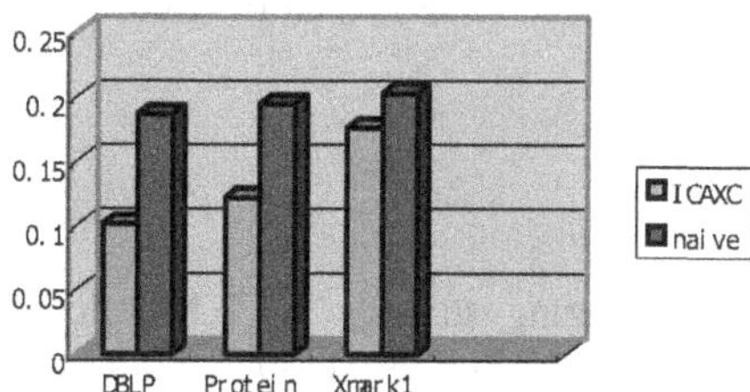


**Fig. 2.** Classified Error Rate of two methods: the ICAXC and naïve method (without dimensionality reduction by ICA)

The first case is to test the accuracy of the ICAXC method. From figure 2, for all the datasets, it is obvious that CER value of ICAXC outperformed that of naïve clustering without ICA method. That's to say, ICA method significantly improves clustering quality. This occurs because the discriminative information in the XML documents is mainly associated with the independent components of the document

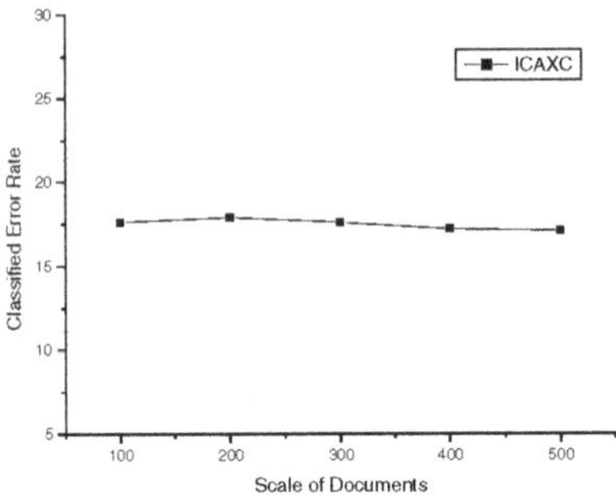


**Fig. 3.** The scalability of the ICAXC method. Xmark datasets: Xmark1, Xmark2, Xmark3, Xmark4 and Xmark5.

matrix D. ICA can mine the projection axes that can be aligned with the data distribution and embody more information. When ICA is used to highlight the discriminative features and at the same time to eliminate ambiguous portions, the performance of the clustering is enhanced.

Note also that the clustering performance of two methods is almost similar in the Xmark1 dataset. This happens due to that ICA technique may not as effective in the sparse matrix as in the normal document matrix.

Then, we test the scalability of ICAXC. In this experiment, Xmark1, Xmark2, Xmark3, Xmark4 and Xmark5 are used as the test dataset, one by one. Figure 3 shows that the Classified Error Rate of these dataset varies very small when the number of the documents increases. It shows that as a dimension reduction technique, the ICA algorithm is a robust and stable algorithm especially when the scale of dataset is large [15]. That is to say, the proposed strategy can be used for a high-volume XML documents collection.

## 5 Conclusions

This paper presents a clustering strategy for XML documents. According to the *D_path*, we introduced the vector representation and distance metric. Then, we apply ICA to reduce the dimensionality of vector space and in the meanwhile, fine the latent features in XML documents. Finally, standard c-means method is used for clustering in reduced Euclidean space. Experimental results show that the method using Independent Component Analysis outperformed the traditional clustering method.

## References

1. Faloutsos C. and Oard D. A survey of information retrieval and filtering methods. Department of Computer Science. University of Maryland, Technical Report, CS-TR-3514, August (1995)
2. A. Nierman and H.V. Jagadish, "Evaluating Structural Similarity in XML Documents," Proc. Fifth Int'l Workshop Web and Databases, June (2002) 1-16

3. Gianni Costa, Giuseppe Manco, Riccardo Ortale, Andrea Tagarelli: A Tree-Based Approach to Clustering XML Documents by Structure. PKDD 2004, Sydney, Australia.(2004) 137-148
4. K. Beyer, J. Goldstein., R. Ramakrishnan., & U. Shaft, "When is the Nearest Neighbour Meaningful?" Proc.of the 7th International Conference on Database Theory, (1999) 217-235
5. L. Parsons, E. Hague, H. Liu, "Subspace clustering for high dimensional data: a review", ACM SIGKDD Explorations Newsletter, Special issue on learning from imbalanced datasets, vol. 6 (1), (2004) 90 - 105
6. Jianghui Liu, Jason TL Wang, Wynne Hsu, Katherine G. Herbert: XML Clustering by Principal Component Analysis. Proc. of ICTAI 2004: 658-662.
7. A. Hyvärinen and E. Oja. "A fast fixed-point algorithm for independent component analysis," Neural Computation, vol, 9, (1997) 1483-1492
8. H.H. Bock, "Probabilistic aspects in clustering analysis," Conceptual and numerical analysis of data, pp., Berlin: Springer-verlag, (1989) 12-44
9. Honkela, T., & Hyvarinen, A. Linguistic feature extraction using independent component analysis. Proc. of IJCNN2004, Budapest, Hungary,(2004)
10. R. Baeza-Yates and B. Ribeiro. Modern Information Retrieval. Addison Wesley, (1999)
11. E. Bingham, A. Kabán, and M. Girolami, "Topic identification in dynamical text by complexity pursuit", Neural Processing Letters, vol. 17(1), (2003) 69-83
12. S. C. Deerwester, S. T. Dumais, T. K. Landauer, G. W.Furnas, and R. A. Harshman. Indexing by latent semantic analysis. Journal of the American Society of Information Science, 41(6), (1990) 391–407
13. Aapo Hyvärinen, Erkki Oja: Independent component analysis: algorithms and applications. Neural Networks 13(4-5) (2000) 411-430
14. T. Kolenda, L. K. Hansen, S. Sigurdsson, Indepedent Components in Text. Advances in Independent Component Analysis, Springer-Verlag,(2000)229-250
15. Tang, B., Shepherd, M., Milios, E. and M.I. Heywood. Comparing and Combining Dimension Reduction Techniques for Efficient Text Clustering.Proc. of International Conference on Data Mining, April 23, Newport Beach, California. 2005
16. DBLP Computer Science Bibliography. 2004. http:// www.informatik.uni-trier.de/~ley/db/
17. Selim, S. Z. And Ismail, M. A. K-means type algorithms: A generalized convergence theorem and characterization of local optimality. IEEE Trans. Pattern Anal. Mach. Intell. 6, (1984) 81–87.
18. Abiteboul, S., Buneman, P., Suciu, D.: Data On The Web: From relations to Semistructured Data and XML. Morgan Kaufmann Publishers, San Francisco, California (2000)
19. Al-Sultan, K. S. and Khan, M. M..Computational experience on four algorithms forthe hard clustering problem. Pattern Recogn. Lett.17, 3, (1996) 295–308
20. S. Zhang, J. T. L.Wang, and K. G. Herbert. Xml query by example. International Journal of Computational Intelligence and Applications, 2(3)(2002) 329–337

# Discovering Multi Terms and Co-hyponymy from XHTML Documents with XTREEM

Marko Brunzel and Myra Spiliopoulou

Otto-von-Guericke-University Magdeburg
{forename.name}@iti.cs.uni-magdeburg.de

**Abstract.** The Semantic Web needs ontologies as an integral component. Current methods for learning and enhancing ontologies, need to be further improved to overcome the knowledge acquisition bottleneck. The identification of concepts and relations with only minimal user interaction is still a challenging objective. Current approaches performed to extract semantics often use association rules or clustering upon regular flat text. In this paper we describe an approach on extracting semantics from Web Document collections which takes advantage of the semi structured content within XHTML (an XML dialect which can be obtained from traditional HTML documents) Web Documents.

The XTREEM (Xhtml TREE Mining) method uses structural information, the mark-up in Web content, as indicators of term boundaries and for co-hyponymy relations.

## 1 Introduction

The realization of the Semantic Web depends on the broad availability of semantic resources, often incorporated in ontologies. Ontology establishment is a process demanding substantial human involvement. To facilitate this demanding process, much research has been devoted to (semi-)automated methods for ontology learning and enhancement. Since semantics are expressed by a lexical layer, such methods must address next to the core task of discovering semantics also the prerequisite task of identifying the terms that represent the concepts [W05]. This terminology issue is still only rarely addressed within ontology learning [BMV01, GTA05].

Many methods tackle this issue by exploiting existing resources such as dictionaries, glossaries or database schemata (e.g. [K99, SSV02]). However, dedicated resources for specific application domains are rare and of low coverage, so that the applicability of such methods is limited. Other methods use plain text as input, converting semi-structured content into plain text [FN99, MS00, BOS05], thereupon eliminating the so-called "syntactic sugar". In this paper, we take the opposite approach: We concentrate on the document structure and use it as guide to the content. Our method XTREEM (XHTML TREE Mining) processes Web sites of XHTML documents and extracts multi-terms and co-hyponyms [COH] by relying solely on page mark-up.

XTREEM has several advantages: It requires minimal human contribution and no linguistic resources. It operates on the syntactic structure, which is independent of

R. Nayak and M.J. Zaki (Eds.): KDXD 2006, LNCS 3915, pp. 22–32, 2006.

national languages and application-specific jargons. It is not constrained by textual borders like sentences and paragraphs and is thus able to find terms that stand in a co-hyponymy relation even if they rarely appear in the same document. XTREEM is thus a complementary method to conventional text analysis, exploiting information that is traditionally skipped, while using the whole of the Web as information source.

The rest of the paper is organized as follows: In section 2 we discuss related work. In section 3 we introduce XTREEM and describe how it processes Web pages, derives vectors of terms by building a feature space of mark-up tags, clusters these vectors on semantic similarity and derives conceptual labels of correlated terms for them. Section 4 contains our first experiments. The last section concludes our study.

## 2 Related Work

A recent overview on Ontology Learning from text has appeared in [BOS05]. Here, we concentrate on methods that consider the Web as information source. Cimiano et al discover hyponymy relations by finding examples of Hearst patterns via the Google API and then analyzing the retrieved documents [CPSS04]. However, they treat documents as plain text, ignoring the semantics implicit in the Web structure.

Web Document structure is used in [E04] to build a knowledge base of extracted entities. Nierman and Jagadish [NJ02] study the structural similarity of XML documents, while Dalamagas et al exploit structural similarities in XML document clustering [DCWS04]. Closer to our work are the studies of Kruschwitz [K01a, K01b], where marked up sections of Web Documents are used to learn a "domain model", because similar mark-up is often used for the representation of similar concepts in Web Documents. Differently from our approach, only local mark-up is exploited: Tag combinations, as reflected in the tree-like structure of (X)HTML documents are not considered. The same holds for the work of Shinzato and Torizawa, who use different tags of HTML documents to find hyponymy relations [ST04]: They consider items of lists but ignoring the role of tag combinations for the representation of semantics.

## 3 The XTREEM Method

We present the XTREEM method for the extraction of semantic relations through the exploitation of Web Document structure. XTREEM is based on mark-up conventions that are present in almost all Web Documents in the HTML (respective XHTML which can be obtained by conversion) format. Authors use different nested tags to structure pieces of information in Web Documents. We find terms that adhere to the same syntactic structure within an XHTML document and apply data mining to find semantically related terms. These desired semantically related pieces of text are not necessarily physically "co-located" i.e. appearing in the same narrow context window as can be seen in the headings example of table1. Both text elements {Wordnet, Germanet} share a common syntactic structure, the series of HTML tags they are placed in. We aim to use such syntactic structures to infer semantic relatedness.

**Table 1.** Semantically related terms, located in different paragraphs or separated by other terms

| Headings, located in different paragraphs | Highlighted keywords, separated by normal text |
|---|---|
| …<h2>Wordnet</h2><br><p>Was developed<br>…</p><br><h2>Germanet</h2><br><p>Analogous<br>…</p>… | … <p> … there are different<br>important standards for building the<br><strong>Semantic Web</strong>. … is<br><strong>RDF</strong>. … <strong>RDFS<br></strong> adds … whereas <strong>OWL<br></strong> is … </p> … |

The tasks of XTREEM are depicted in Fig. 2 and described in Section 3.2. Before doing so, we introduce some basic terminology in Section 3.1.

### 3.1 Web Documents

*Web Document D:* A Web Document (Web page) is a semi-structured document following the W3C XHTML standard. XHTML is a XML dialect, wherein the former HTML standard has been adopted to meet the XML requirements. Traditional legacy HTML documents are converted to XHTML documents, as it is performed by all popular Web browsers too. The major constituents of XHTML documents are tags (mark-up elements) which enclose text (text elements) as described in the following. In the XML terminology only the terms "element" and "text" are used, but for audibility we will use "mark-up element" and "text element" in the following.

*Text Element T:* A "Text Element" within a Web Document is a continuous span of text without tags; tags form its border. It can be either (1) a single token without any white space like "Wordnet" in line 8 of Fig. 1, (2) a multi-token term like "Lexical Resources" in line 6 of Fig. 1 or (3) a long sequence of tokens like the texts surrounded by paragraph tags in the same Figure. For our objectives, we are interested in identifying text elements of the first two types: co-hyponyms can be single or multi-token terms. As we will see in the next subsection, XTREEM skips text elements that occur rarely in the collection, so that texts of the third type are filtered out anyway.

```
1  <html>
2  <html><head>
3  <html><head>…
4  <html></head>
5  <html><body>
6  <html><body><h1>Lexical Resources …</h1>
7  <html><body><p>…</p>
8  <html><body><h2>Wordnet</h2>
9  <html><body><p>Was developed …</p>
10 <html><body><h2>Germanet</h2>
11 <html><body><p>Analogous to Wordnet for the English …</p>
12 <html><body>…
13 <html></body>
14 </html>
```

**Fig. 1.** Document Paths for Text Elements in a XHTML Tree

*Mark-up Element:* According to the XHTML standard, a "Mark-up Element" is a fixed set of tags which can be used to structure XHTML documents. These tags are interpreted by Web browsers during document rendering.

*Document Path P:* For each text element a document path, defined as the sequence of mark-up elements from the document root to the text element within the XHTML tree, can be constructed. For example, the heading "Wordnet" in line 8 of Fig. 2 has the document path `<html><body><h2>`.

Document Path = [Mark-up Element Name]*

## 3.2 The XTREEM Procedure

The XTREEM discovers multi-terms and co-hyponyms for a domain of discourse by mining Web Documents. The XTREEM process encompasses the tasks depicted in Fig. 2. Those tasks extend the conventional process of text mining by a task that builds the text collection itself from the Web. The core of XTREEM are the parallel tasks for Building the Feature Space and Building the Data Space. Briefly, the feature space consists of text elements, while the data space consists of document paths leading to the text elements, i.e. to the features. The tasks of XTREEM are described below.

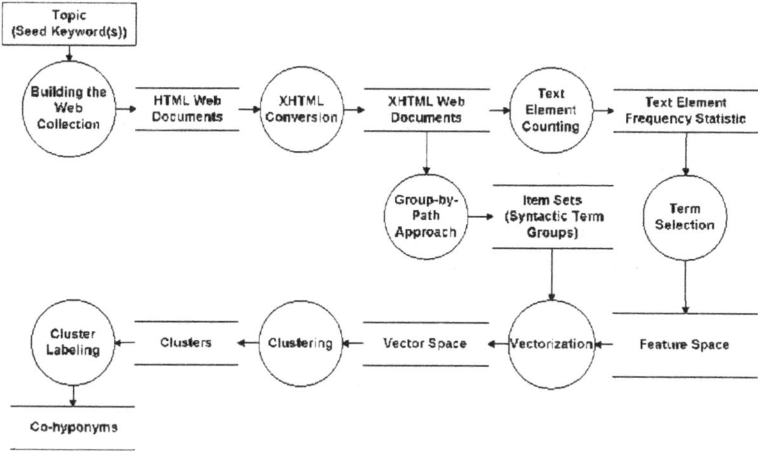


**Fig. 2.** Data-flow Diagram of the XTREEM procedure

**Building the Web Collection:** The input to XTREEM is a small set of keywords, the "seed", which characterizes the target domain. Rather than expecting a well-prepared collection of appropriate documents, XTREEM collects documents from the Web by invoking a crawler or by retrieving document references from internet search engine web services.

Hence, the user input to XTREEM is limited to specifying a seed that describes the domain of discourse adequately and guarantees broad coverage. Example seeds may be (1) "Semantic Web" for the Semantic Web, (2) "tourism" for tourism or (3) "cardiology" for everything associated with heart medicine. More specific terms that characterize the domain, such as "ontology" or "XML" for the Semantic Web or "hotel" for "tourism" are possible but not necessary.

**XHTML Conversion:** This simple task transforms Web Documents complying to the older HTML standard in XHTML. Moreover, the converter eliminates some existing format errors, thus dealing with malformed Web Documents as well.

**Text Element Counting:** We create a frequency statistic on all Text Elements. For efficiency, a threshold on the maximum length of text elements can be incorporated to refuse long sequences of text at an early stage. The longer a text element is, the more unlikely that it is a term.

**Term Selection:** For the feature space construction, the human expert should specify the desired number of features as value of the threshold n. Small values of n are more appropriate if the expert is interested to learn the base terminology for the domain, while large values are more reasonable if the goal is to collect as many terms and multi-terms as possible and acquire co-hyponyms for them.

Due to the low frequency, long text elements (text which is not marked up) have nearly no chance to get into the feature space, while short terms which consist of more than one token and which are used frequently inside the document collection get into the feature space. This has the positive effect, that our approach has an implicit multi term recognition, which otherwise would be a complex Natural Language Processing problem of its own, e.g. the multi token terms "data mining", "Semantic Web" and "Resource Description Format" are recognized by this approach. Web Document specific words such as home, `contact`, `back`, `top`, `site_map` are rejected with help of a domain neutral Web content stopword list.

**Group-by-Path Approach:** The XHTML tree is traversed by the XTREEM algorithm, for each encountered text element the document path is built. Document path and the text element are stored together for later processing. When the whole document is traversed, we group text elements that have the same document path as its predecessor. E.g. in our example (Fig. 2), Wordnet and Germanet both have `<html><body><h2>` as document path, and, thus become members of the same set of terms `{Wordnet, Germanet}`. Usually, authors use different tags and therefore things separate according to different tags, resulting in different documents paths, therefore several Text Element Sets stemming from one document are possible.

---

**Algorithm 1.** The XTREEM Group-By-Path approach on a XHTML document

---

Input: D
Output: n 'sets of T'

1: for all T in D: create the corresponding P → store P associated with T
2: create the set of n unique P
3: for all n unique P:
    for all T: which T are associated with P → store T
    store set of T
**return** n 'sets of T'

---

The resulting sets are filtered: only sets with cardinality greater than min and cardinality smaller than max are further processed. This corresponds to the usage of only those mark-up structures, which are regarded as providing a useful separation. Here precision is preferred over recall.

Next we will contrast how this approach is different to traditional processing of documents.

| Traditional processing | XTREEM processing |
|---|---|
| If a page contains the text elements `{Contact, Map, Back, Lexical Resources, Wordnet, Germanet}`, one would regard all this terms as a set and model the document as a vector over those terms. | XTREEM processing: According to our approach, which incorporates the structure of the XHTML tree, it is more likely that the text elements form more homogenous term sets, e.g. the 4 term sets `{contact, map}`, `{back}`, `{lexical structures}` and `{Wordnet, Germanet}`. XTREEM groups text elements with the same document path together, thus resulting in more homogenous instances which facilitate further processing to reveal semantic relations among text elements. |

Note that we use element tags only to infer siblingness of elements. We do not consider the meaning of the tags.

The term sets found by this approach can be used for different purposes. In the following we will describe the application of clustering upon these term sets with the goal to eliminate terms which do not belong in such sets, because the semantic relation is of another type than typical inside a set or because there is no semantic relation at all among the set members.

**Vectorization:** The term sets obtained by the Group-by-Path procedure in step 3 are now vectorized according to the feature space build in step 4. We only process term sets with more than one unique member (for our purpose, finding semantic relatedness, a single term is not useful because for the desired semantic relations at least 2 terms are necessary). Each term set (text element set, transaction) is used to form an instance (vector, record, matrix row). Afterwards, TF-IDF weighting is performed, where IDF refers to the number of vectors, i.e. document paths, rather than to the number of original documents

| | Wordnet | Germanet | Euroword net | Semantic Web | ... |
|---|---|---|---|---|---|
| DocumentA<html><body><h2> | 1 | 1 | 0 | 0 | ... |
| DocumentB<html><body><table><h1> | 1 | 0 | 1 | 0 | ... |
| DocumentC<html><body><p>... | 0 | 0 | 0 | 1 | ... |
| ... | ... | ... | ... | ... | ... |

**Fig. 3.** Exemplary fragment of a Vectorization

**Clustering:** The objective of the clustering task is the discovery of correlated features, more precisely of co-hyponyms. The Vectorization obtained in the prior step has the tendency to reveal semantic related terms. One way to get these related terms is the application of a clustering algorithm. Association Rules Mining would be an alternative method. For clustering a K-Means algorithm with cosine distance function was applied.

The amount of clusters to be generated can be set on the algorithm. The clustering algorithm creates clusters of instances, which are not useful on our objectives themselves. The desired result (related terms) has to be obtained by the following post processing step.

**Cluster Labelling:** As we are not directly interested in whether documents paths (with their associated terms) fall into a cluster, we want to see semantic relatedness, expressed through the characteristics of clusters. A "label" is a subset of the features supported by the cluster members, such as the m most frequent features or the features with higher support than a threshold. According to our objectives, these features are semantically correlated, since they appear together in many instances.

## 4 Experiments

We present here our first preliminary experiments on the discovery of multi-terms and co-hyponyms with XTREEM. The evaluation of an agnostic method like XTREEM is intriguing for the following reasons: First, the establishment of the Web Document collection for a given seed of keywords is part of the XTREEM procedure; hence, we cannot compare with a method that is applied on a well-prepared corpus. Second, only a human expert can decide whether a multi-token object is indeed a multi-term and whether two features are in co-hyponymy relation within an arbitrary domain of discourse. In future work, we intend to test XTREEM against the multi-terms and co-hyponyms of a given ontology, using it as gold standard for a given domain of discourse. In this study, we concentrate on showing the potential of XTREEM in proposing multi-terms and co-hyponymy candidates for the exemplary domain of discourse "Semantic Web, Ontology". For comparison purposes, we have devised a simple agnostic method that discovers correlated features by analyzing the plain text.

### 4.1 The Web Document Collection

The establishment of the document collection is the first task of the XTREEM procedure. The seed consisted of the keywords "Semantic Web" and "Ontology". We used Google API for retrieving. Under standard settings, Google returns a maximum of 1000 documents per query. To increase the coverage, we have issued for each keyword K in the seed several queries containing the seed and one additional constraint, namely asking for (1) htm documents, (2) html documents, (3) excluding ps and pdf documents and (4) excluding all of the above, so that e.g. php documents could be retrieved. We have thus acquired 4 sets of Web Documents for each keyword. We merged those sets for all keywords, eliminating duplicate documents. The result was a set of 4209 distinct URLs, from which we retrieved 4015 Web

Documents from 2112 domains. From these, we have removed approximatly 10 percent documents that were recognized as non-English language documents.

### 4.2 Experiment 1 - XTREEM

According to the preprocessing tasks of XTREEM, the Web Documents have been converted to XHTML and the frequencies of text elements over the whole document collections have been counted. We have chosen the 1000 most frequent text elements as features. The Group-by-Path algorithm has processed 49365 document paths, using the threshold values min=1 and max=+infinity. The threshold m on the number of non-zero values per vector was set to 2, so that 6109 vectors were retained.

The vectors have been weighted using TF-IDF and the K-Means clustering algorithm has been applied, setting K=100. We refer to these results as "document path clusters" or "path clusters" for short. Then, each cluster was labeled by its k=10 most frequent features. In Table 2 we show the features in the labels of a selection of three clusters. These clusters were selected because the correlated features in their labels were the easy to interpret. However, many further clusters contained no less informative labels. As can be seen from the table, the cluster labels are quite intuitive. The rightmost one contains 9 publishers where books, journals or articles on the domain of discourse have appeared. The middle cluster contains names of researchers; the two forms of the forename of the last person are remarkable here. The left cluster contains 9 key terms associated with the Semantic Web and with ontologies. Next to the fact that all those terms are related to the domain of discourse, the clear thematic separation of the clusters must be stressed.

**Table 2.** Clusters of Document Paths (characterized by 10 most frequent features)

| | | |
|---|---|---|
| ontology | tim_berners_lee | springer |
| taxonomy | deborah_l_mcguinness | wiley |
| thesaurus | eric_miller | acm |
| source | ora_lassila | elsevier |
| controlled_vocabulary | stefan_decker | ieee |
| metadata | brian_mcbride | march_april |
| topic_maps | dan_brickley | mit_press |
| concept | j_r_me_euzenat | springer_verlag |
| faceted_classification | jim_hendler | computing |
| is_a | james_hendler | ieee_computer_society |

### 4.3 Experiment 2 - Application of Conventional Procedure

For comparison purposes, we have designed a conventional text analysis method that has prepared, vectorized and clustered the Web Documents as plain texts. We have used similar constraints: For the feature space, we have selected the 1000 most frequent features. Vectors with less than m=2 non-zero values were removed, resulting in 3089 out of 3829 vectors as input to the clustering algorithm. Again, the K-means with cosine similarity was used, setting K=100. Each of the 100 clusters, hereafter denoted as "document clusters" was labeled with the k=10 most frequent features in it. The labels of four clusters are shown in Table 3; again, these are the clusters whose labels can be most easily interpreted.

As can be seen, those labels are much more diffuse: The same feature appears in many labels, terms characteristic for the domain are mixed with generic words (e.g. entity and introduction in the second cluster to the right), while the few recognized names of researchers appear together with names of institutions and with some generic names (department of computer science, chair).

We have experimented with this method for larger values of K as well. If K is between 300 and 500, then some homogeneous clusters of similar label quality to those of XTREEM can be found. However, this implies that the human expert must study a much larger number of less interesting clusters to identify reasonable good labels.

**Table 3.** Clusters of Documents (characterized by 10 most frequent features)

| | | |
|---|---|---|
| ontoedit | department_of_computer_science | ontology |
| rdf | university_of_maryland | relation |
| oil | agents_and_the_semantic_web | abstract |
| semantic_web | james_hendler | attribute |
| daml | darpa | conclusion |
| ontolingua | chair | entity |
| project | hendler_cs_umd_edu | introduction |
| semtalk | ian_horrocks | knowledge_base |
| protege | nature | semantic_web |
| tool | semantic_web_services | description_logic |

### 4.4 Comparison of the Findings

The differences between the document clusters of the conventional method and the path clusters of XTREEM can be summarized as follows:

- Document clusters are more diffuse, containing features related by arbitrary kinds of semantic relationships.
- The semantic relationships among the features in each path cluster are easily recognizable. This is indicated by the fact that a summarizing concept can be assigned to each of these clusters, serving as parent concept. Hence, the semantic relationship is a sibling-relationship – co-hyponymy: For example, the clusters in Table 2 refer to (1) instruments for the representation of meta-data types, (2) to persons and (3) to publishers.

A posteriori, the supremacy of XTREEM towards simple text analysis is not astonishing: When authors group texts at the same level into itemlists, headlines etc, they are usually motivated by the intention to present sibling concepts in an intuitive way.

For the path-clusters, a human expert can often easily name the implicit but unnamed parent concept and filter out the erroneous terms of the cluster. This requires much less effort than the manual identification of co-hyponyms from groups of loosely correlated features.

The terms in the traditional document-clusters are not semantically unrelated, but the relations are manifold and can bot be eaysls named.

## 5 Conclusions and Future Work

We have presented XTREEM, an agnostic method for the discovery of semantic relations among terms on the basis of structural conventions in Web Documents. We exploit the interplay of structure and content in Web Documents to find groups of terms which have a certain syntactic structure within a Web Document in common.

Our first results indicate that rerms appearing in the same cluster, i.e. co-occuring in different documents with the same mark-up grouping are good co-hyponymy candidates.

Our method is only a first step on the exploitation of the structural conventions in the Web for the discovery of semantic relations. We will next perform an evaluation of of the extracted terms and co-hyponomy relations. Discovering the coresponding hypernym for the co-hyponyms is a further desireable extension. In our future work we also want to investigate the impact individual mark-up element tags.

## References


[BMV01] R. Basili, M. Missikoff, and P. Velardi, Identification of relevant terms to support the construction of Domain Ontologies, ACL-0 1 workshop on Human language Technologies, Toulouse, France, July 2001

[BOS05] P. Buitelaar, D. Olejnik, M. Sintek , Ontology Learning from Text: Methods, Evaluation and Applications, Frontiers in Artificial Intelligence and Applications Series Volume 123, IOS Press, Amsterdam, 2005

[COH] http://www.websters-online-dictionary.org/definition/english/co/co-hyponyms.html

[DCWS04] Dalamagas, T. & Cheng, T. & Winkel, K.-J. & Sellis, T. (2004). A Methodology for Clustering XML Documents by Structure. Information Systems. In press.

[E04] O. Etzioni, M. Cafarella, D. Downey, S. Kok, A.-M. Popescu, T. Shaked, S. Soderland, D. S. Weld, A. Yates. Web-Scale Information Extraction in KnowItAll. Proceedings of the 13th International WWW Conference, New York, 2004.

[FN99] D. Faure, C. Nedellec. Knowledge acquisition of predicate argument structures from technical texts using machine learning: the system ASIUM. EKAW '99, volume 1621 of LNCS, pp 329-334.

[GTA05] L. Gillam and M. Tariq and K. Ahmad, Terminology and the Construction of Ontology. Terminology 11 2005 , pp55-81. John Benjamins Publishing Company.

[K01a] Kruschwitz, U. "A Rapidly Acquired Domain Model Derived from Mark-up Structure". In Proceedings of the ESSLLI'01 Workshop on Semantic Knowledge Acquisition and Categorization, Helsinki, 2001.

[K01b] U. Kruschwitz. Exploiting Structure for Intelligent Web Search. Proc. of the 34th Hawaii International Conference on System Sciences (HICSS), Maui Hawaii 2001, IEEE

[K99] V. Kashyap. Design and creation of ontologies for environmental information retrieval. Proc. of the 12th Workshop on Knowledge Acquisition, Modeling and Management. Alberta, Canada. 1999.

[MS00] A. Maedche and S. Staab. Discovering conceptual relations from text. In Proc. of ECAI-2000, pp. 321-325.

[NJ02] Nierman, A. & Jagadish, H.V. (2002). Evaluating Structural Similarity in XML Documents. In Proc. of International Workshop on the Web and Databases, 61-66.

[SSV02] L. Stojanovic, N. Stojanovic, R.Volz. Migrating data-intensive Web Sites into the Semantic Web. Proc. of the 17th ACM symposium on applied computing. ACM press, 2002. 1100-1107.

[ST04] K. Shinzato and K. Torisawa. Acquiring hyponymy relations from web documents. In Proceedings of the 2004 Human Language Technology Conference (HLT-NAACL-04), pages 73--80, Boston, Massachusetts, 2004.

[W05] H.F. Witschel. Terminology extraction and automatic indexing - comparison and qualitative evaluation of methods. In Proc. of Terminology and Knowledge Engineering (TKE), 2005.

# Classification of XSLT-Generated Web Documents with Support Vector Machines

Atakan Kurt and Engin Tozal

Fatih University, Computer Eng. Dept.,
Istanbul, Turkey
{akurt, engintozal}@fatih.edu.tr

**Abstract.** XSLT is a transformation language mainly used for converting XML documents to HTML or other formats. Due to its simplicity and flexibility XML has replaced traditional EDI file formats. Most e-business applications store data in XML, convert XML into HTML using XSTL, and publish the HTML documents to the web. In this paper we argue that the use of XSLT presents an opportunity rather than a challenge to web document classification. We show that it is possible to combine the advantages of both HTML and XML into classification of documents at the XSLT transformation stage, named *XSLT classification*, to attain higher classification rates using Support Vector Machines (SVM). The results are both expected and promising. We believe that XSLT classification can become a favorable classification method over HTML or XML classification where XSLT stylesheets are available.

## 1 Introduction

Data mining has been applied to a much wider spectrum of application domains including web, GIS, multimedia in the last decade facilitated by many remarkable advancements in various branches of information and computing technologies. XML (eXtensible Markup Language, http://www.w3.org/XML/) is certainly one of those information technologies that have dramatically impacted many application domains. By bringing structure to unstructured documents, XML practically became a synonym for semi-structured documents in the area of digital libraries or information retrieval.

The widespread use of XML in e-business applications has resulted in the definition of many domain specific XML vocabularies such as ebXML, VRML, SVG. More recently applications that produce dynamic or static HTML documents have started generating documents/data in XML format first, then converting the XML documents to medium or client specific formats including HTML, XML, text, PDF(Portable Document Format), WML (Wireless Markup Language), etc using XSLT (eXtensible Style Language Transformations, http://www.w3.org/TR/xslt), thus simplifying overall software engineering process and cutting down development costs among other obvious advantages.

From the data mining point of view, this new XML-to-HTML-by-XSLT trend seems to complicate things at first look, as the effects of such transformations on the classification of web documents whether HTML or XML were not considered before. A number of web page or site classifcation techniques based on HTML have been

R. Nayak and M.J. Zaki (Eds.): KDXD 2006, LNCS 3915, pp. 33 – 42, 2006.

introduced into web mining literature [9, 10]. Different approaches are taken in these studies. Text-only approach removes all markups and performs classification based on pure content [7]. HyperText approach considers the markup tags to assign weights of features [8, 9]. Link Analysis approach builds its classification model links among web pages [11, 12]. Semi-Structured Document or XML classification techniques not only consider the textual content or the text but also the markup. Structured vector model [13], tags each word with the enclosing markup to generate a feature set. A different semi-structured document classifier [14] proposes a model based on Bayesian Networks in which the training is done for sub-sections of documents.

In this paper we show that XSLT presents unique opportunities rather than new challenges in web classification. The idea is to combine advantages of HTML and XML with the power of XSL transformations which is presented in more detail in Section 2. The experiments performed on a small data set reveal that XSLT transformation outperforms both HTML and XML classifications using Support Vector Machines, confirming similar findings performed in a previous study [1, 2] using Naïve Bayes.

This paper is organized as follows: In Section 2 we define XSLT and the motivation behind the use of XSLT in web applications. The web/document classification framework based on XSLT used to perform the experiments is briefly introduced in Section 3. We discuss the data set, the experimental setup and a short evaluation of the results in Section 4. The conclusions are drawn in Section 5.

## 2 Background and Motivation

XSLT is a part of XSL (Extensible Stylesheet Language) standard that is used to define a set of transformation rules for converting XML documents into different formats as shown in Fig. 1. In a way, XSLT is to XML as CSS (Cascading Style Sheets) is to HTML. However XSLT is much more powerful than CSS as it can be used as a full-featured programming language. XSLT Stylesheets (transformation programs written in XSL), as they are called, are themselves written in a specific XML format defined by a DTD (Document Type Definition), and therefore can be processed by other XSLT stylesheets as data. Different stylesheets can be used for different requirements. For example, one stylesheet can be employed to produce WML output for WAP (Wireless Application Protocol) enabled cell phones, a different stylesheet for printer output, and another one for handheld computers with small displays.

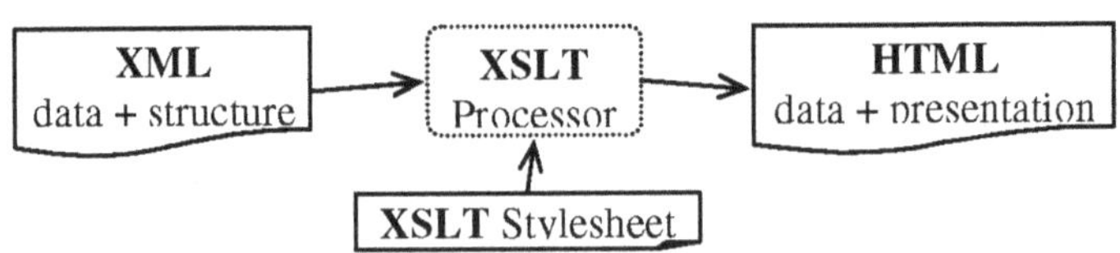


**Fig. 1.** The Extensible Stylesheet Language Transformations

As shown in the figure, XML enables us to define the structure and separate the data/content from presentation which is mostly medium/application specific. This advantage over HTML is exploited in the web classification framework used in this study. XSLT stylesheets are executed by XSLT processors such as Saxon, and Xalan that are available as APIs or in mainstream web browsers. XSLT stylesheets can be used in various combinations. Two such scenarios are shown in Fig. 2.

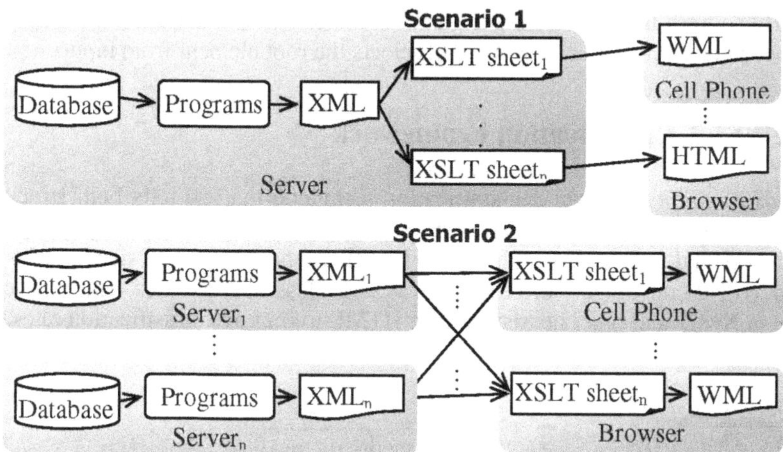


**Fig. 2.** The use of XSLT in Web Applications

XML

```
<source id="10021">
<type>article</type>
<status>APPROVED</status>
<title>XSL</title>
<author><fname>John</fname>
<lname>Sabre</lname>
</author>
<comment>Appropriate for
publishing</comment>
</source>
```

XSLT

```
<xsl:stylesheet version = '1.0'
     xmlns:xsl='http://../Transform'>
<xsl:template match="/">
<html><head><meta name="keywords"
content="article, xml, xslt,
xsl"/><title>XSL
Era</title></head><body>
<h1><xsl:value-of
select="//title"/></h1>
<h2><xsl:value-of
select="//author/fname"/> <xsl:value-
of select="//author/lname"/></h2>
Articles about XML, XSL, and new web
technologies…
</body></html>
</xsl:template>
</xsl:stylesheet>
```

HTML

```
<html><head>
<meta name="keywords"
content="article, xml,
xslt, xsl, web"/>
<title>XSL ERA</title>
</head><body><h1>XSL</h1>
<h2>John Sabre</h2>
Articles about XML, XSL,
and new web technologies…
</body></html>
```

**Fig. 3.** The Sample XML, XSLT and HTML Documents

Fig. 3 contains a sample XML document, an XSLT stylesheet applied to this sample document, and the produced HTML output. XSL commands and templates are indicated with the *xsl*: namespace in the stylesheet. XSLT stylesheets are composed of *templates* to be *match*ed from the input. All literals and the results of applying the templates are copied to the output. *Values-of* elements from the input can be *select*ed. Looping, conditional processing, functions are available. Built in filtering and sorting capabilities are parts of the language as well. XPath, XML Path Language is used to selectively address the parts of input XML document. For example *//title* selects all *<title>* elements, while special '/' symbol selects the root element from input.

## 3 The XSLT Classification Framework

*XSLT classification* is a hybrid classification technique that exploits both structural markup in XML and presentational markup features in HTML. The logical system layout is depicted in Fig. 4. In this figure HTML is shown in gray, since we do not work on HTML documents directly. The XSLT classification uses data (content), structure in XML, and other heuristics from HTML to improve classification rates.

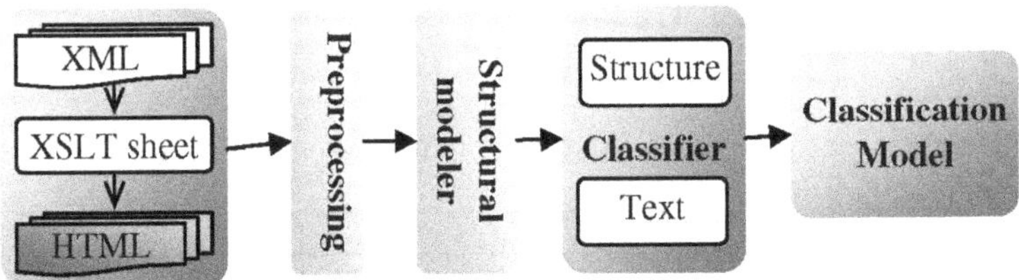


**Fig. 4.** The XSLT Classification Overview

We think that XSLT classification produces better results than HTML classification, because XML documents contain structural information of nesting and content specific markup vocabulary not present in HTML.

We believe that XSLT classification is better than XML classification too, because:

- An XML document usually contains meta-data that is not related to the subject of the document but used for other purposes by the generators of document. Usually that data is not presented to end-user, and should be omitted in the classification process. Elements such as *id, type, status* are examples of meta-data in Fig. 3. We can eliminate meta portions of XML document using XSLT classification
- There may be some literals valuable in classification presented to the user as part of HTML but not a part of the XML document such as sentence *"Articles about XML, XSL, and new web technologies..."* as shown in Fig. 3. This type of data can be processed with XSLT easily.
- Some important data that is embedded in HTML tags like <title>, <img>, <meta> but not present in XML should be captured with XSLT.

- Sometimes only a part of an XML document may be relevant for the end-user or relevant to the subject of the HTML document. In this case, the irrelevant parts of the XML should be discarded all together in the classification.
- It may be the case that an HTML output could be combination of a set XML documents. It would be wiser to consider the output HTML, instead of input documents in classification which is the case with XSLT classification.

The architecture of web classification framework based on XSLT is shown in Fig. 5. The framework consists of three modules; *Preprocessor, Semi-Structured Document Modeler, and Classifiers*. The system accepts a set of XML documents and an XSLT stylesheet to transform them in HTML or other formats as input. The output of the framework is a classification model for the given training set using support vector machines. The framework can be used to classify documents in both Scenario 1 and Scenario 2.

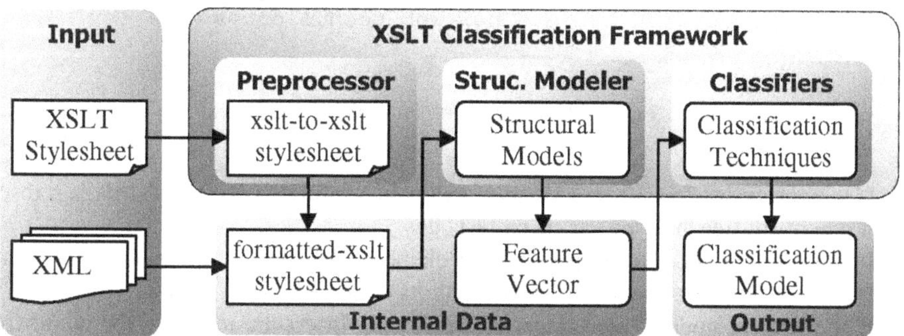


**Fig. 5.** The XSLT Classification Framework Architecture

The original XSLT document is passed to the preprocessor in which an XSLT-to-XSLT stylesheet is applied to produce a new XSLT document called *formatted-XSLT stylesheet*. The formatted-XSLT is an XSLT stylesheet that contains transformation rules in the original XSLT, at the same time, is able to produce appropriate outputs of XML documents for the structural modeler. These *formatted-XML* documents (not shown in the figure) generated by applying formatted-XSLT to the XML documents are then fed to the structural modeler which applies one of the structural models defined in the study [1]. The modeler produces a feature vector for the term frequency vectors (not shown) used later in the classification step. A feature vector usually contains all unique words mostly prefixed with the XML tags in which they reside in the original XML document. Term frequency vectors are generated for all formatted-XML documents and passed as a training set to the classifier. The classifier contains a number of classification algorithms in Weka. SVM [5] is used in this study, as it was reported in [3,4,6] to perform better that many others on text. The classifier creates a model based on the training set. This model is used to classify new documents. In the following sub-sections we briefly discuss components of the framework.

### 3.1 Preprocessor

In the preprocessing step, an XSLT-to-XSLT stylesheet is applied to the original XSLT stylesheet to generate *formatted-XSLT stylesheet*. The XSLT-to-XSLT stylesheet simply traverses each element of the original XSLT document and does the following to produce the XSLT stylesheet which is referred to as result tree below. The *ancestor-or-self* refers to creating a string by concatenating all the element names from the root to the innermost element separated with '-_-' as shown in Fig. 6.

```
<?xml version="1.0" encoding="UTF-8"?>
<document-root>
article xml xslt xsl XSL Era <source-_-title>XSL</source-_-
title>
    <source-_-author-_-fname>John</source-_-author-_-fname>
    <source-_-author-_-lname>Sabre</source-_-author-_-lname>
Articles about XML XSL and new web technologies
</document-root>
```

**Fig. 6.** The Sample Formatted-XML Document

- If the current node is an *xsl:element* node whose name is not an HTML tag, insert it into result tree and process its child-nodes.
- If the current node is an *xsl:vlaue-of* element, then insert the value referred by *select* attribute into the result tree with all its *ancestor-or-self* hierarchy.
- If the current node is an *xsl:text* element, then insert the text into the result tree with all its *ancestor-or-self* hierarchy.
- If the current node is any other XSLT element -*xsl:variable, xsl:param, xsl:with-param, xsl:if, xsl:when, xsl:choose, xsl:otherwise, xsl:copy, xsl:copy-of, xsl:sort, xsl:for-each*- put it directly into the result tree and process its children.
- If the the node is an HTML *title, meta, img/alt* tag, insert its content into the tree.
- If any other string literals, copy them into the result tree.

### 3.2 Semi-structured Document Modeler

Semi-Structured document modeler generates a feature vector from the set of formatted-XML documents. Later the frequencies of each unique feature or word are placed into term frequency vectors for each formatted XML document using a structural representation model.

There are a number of alternatives, as explained in [1], of incorporating structural information into document classification such as prefixing the word with the innermost enclosing tag or all inclosing tags etc. The strength of each alternative model is affected both by how the structure is represented in the term frequency vectors and by the variations in element, removal or insertion of inter elements, and the swap of elements in the document. We show how a document is represented in feature vectors in this study. The example below is based on XML document given in Fig. 3. `source, author, fname` are tags, *John* is a text content.

**Table 1.** A Sample Term Frequency Vector

| Feature Vector | Term Frequency Vectors | | | |
|---|---|---|---|---|
| | $D_1$ | $D_2$ | ... | $D_n$ |
| ... | | | | |
| source | 8 | 6 | | 4 |
| author | 4 | 2 | | 0 |
| source.author | 3 | 1 | | 0 |
| fname | 3 | 1 | | 1 |
| source.fname | 2 | 2 | | 2 |
| author.fname | 2 | 3 | | 0 |
| *john* | *1* | 0 | | *1* |
| source.*john* | 1 | 0 | | 1 |
| author.*john* | 1 | 0 | | 0 |
| source.*john* | 1 | 0 | | 1 |
| . . . | | | | |

A feature for term frequency vector is created as follows: each unique word or element is a feature. Each word and element is prefixed with each of its ancestor separately to create new features. Furthermore each of the ancestor elements is prefixed with their ancestor elements to create new features. Table 1 shows term frequency vectors for the "*<source-_-author-_-fname>John</source-_-author-_-fname>*" fragment of formatted-XML document given in Fig. 6. Values shown in the table are not actual values.

Although the structure is captured in a loose manner (i.e. we do not capture ancestor hierarchy in a strict manner), the complete document hierarchy is captured.

# 4 Experiments, Results, and Evaluation

We performed a set of experiments to compare the classification rates for HTML, XML, and XSLT classifications using the framework described above. The framework was implemented in Java and XSLT. We used Weka for the classification algorithms. 2/3 of documents are used for training and 1/3 for testing with 10-fold cross validation to improve the reliability. The element/attribute names and the words in text are stemmed (taking the root of a word) in all experiments, as it is a common practice in text mining.

## 4.2 Dataset

Current dataset repositories on the web do not provide a proper dataset for the XSLT classification. We generated XML/XSLT version of web pages from 20 different sites belonging to 4 different categories; *Automotive, Movie, Software, News & Reference.* The sites in *News & Reference* category contain news and articles about movies, automobiles and software health and literature. This should result in a more difficult classification task. The data set can be downloaded freely from http://www.fatih.edu.tr/~engin. 100 XML documents were generated from the web sites. These documents are evenly distributed among categories. XML documents are created in a way that they have various structures, element and attribute names, and nesting to mimic that they are generated by different people/applications. An XSLT stylesheet producing exactly the same presentation with all links, images, embedded objects, literal strings and non-printable data like meta, style, script tags etc. of actual HTML page is generated for each web site. When the XSLT is applied to the XML document, it combines static content with dynamic content. Static content includes menus, scripts, headings, and other similar material. Dynamic content is the data retrieved from the XML files. By combining the two contents, an HTML or XHTML page is created.

**Table 2.** TFV Sizes

| Data Type | # Features |
|---|---|
| XML | ~37000 |
| HTML | ~7000 |
| XSLT | ~34000 |

**Table 3.** Dataset Properties

| Data Set | # Classes | # Sites | # Documents |
|---|---|---|---|
| 1 | 4 | 17 | 91 |
| 2 | 4 | 14 | 75 |
| 3 | 3 | 13 | 70 |
| 4 | 3 | 13 | 60 |

Since data size is quite limited, we have created 4 different versions of the data set by excluding either some categories or some web sites or both randomly from the original data set. Characteristics of data sets are shown in Table 2 and Table 3. We conducted experiments on these data set using SVM and Naive Bayes [15,16,17] to compare the results with the previous study [2] which used only the Naïve Bayes. Experience dictates that not all classification algorithms do well with a certain type of data. Data mining is an experimental science. It is also a common practice, called voting, to apply a number of different techniques to classify a new instance by choosing the category preferred by the highest number of techniques. For these reasons, we think that it is necessary to conduct further experiments with different techniques and datasets on XSLT classification as explained in the next section.

All presentation markups are removed while generating feature set for HTML documents. Yet contents of *meta, title, anchor* and alternative name for *img* tags are included into the feature set. The structural modeling technique explained in Section 3.2 is not only used to generate feature set for XSLT documents but also for XML documents. Moreover, %2 to %4 noise is introduced into XML documents instead of using actual XML meta data.

### 4.4 Results and Evaluation

The experimental results are shown in tabular form per data set in Table 4 and average classifications over all datasets are depicted in Fig. 7. In general XSLT classification yields considerably higher accuracy rates than both HTML and XML classification, while XML classification produced slightly better accuracy rates than HTML classification. The results reveal that both SVM and Naïve Bayes deliver similar rates confirming each other's output.

In all data sets except the last one, the XSLT classification performs better than the other two, while XML classification yields better scores than HTML classification.

Noticeably Naïve Bayes outperforms SVM on both XSLT and XML data. Since the numbers are close and data sets are small, between Bayes and SVM it s not possible determine which one is better. However there seems to be marked difference between the two methods on HTML data. As shown in Table 2, representing structure in XML and XSLT classifications results in much larger term frequency vectors than those of HTML classification. Since XML and XSLT produce higher classification rates, this is a trade-off between accuracy and space. However an empirical threshold value can be used to reduce the term frequency vector size in XML and XSLT classifications. As shown in Table 4 SVM takes longer in building classification model compared to Naïve Bayes (NB).

**Table 4.** Accuracy Rates of SVM and NB

| Data Set | Data Type | Bayes % | SVM % | Bayes Time | SVM Time |
|---|---|---|---|---|---|
| 1 | HTML | 94.44 | 96.6 | 20 | 490 |
| | XML | 97.7 | 97.7 | 60 | 1000 |
| | XSLT | 100 | 100 | 50 | 950 |
| 2 | HTML | 93 | 96 | 30 | 2090 |
| | XML | 97.3 | 97.3 | 910 | 890 |
| | XSLT | 100 | 100 | 50 | 1630 |
| 3 | HTML | 92.8 | 95 | 20 | 360 |
| | XML | 97.14 | 95.7 | 50 | 660 |
| | XSLT | 100 | 100 | 50 | 590 |
| 4 | HTML | 95 | 95 | 20 | 270 |
| | XML | 95 | 95 | 40 | 880 |
| | XSLT | 100 | 96.6 | 50 | 890 |

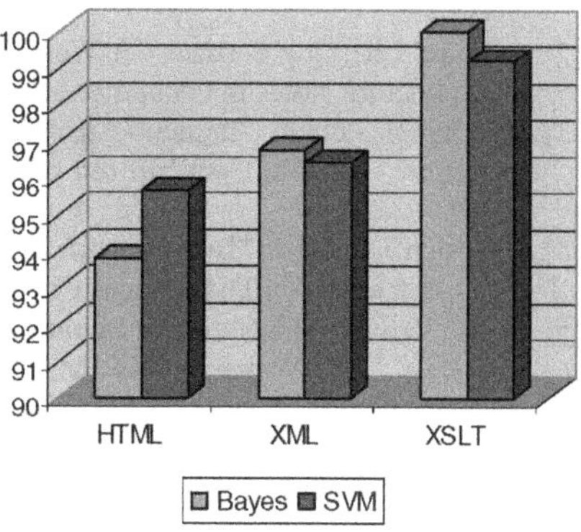


**Fig. 7.** Average Accuracy Rates

## 5 Conclusions

XSLT is used in more and more applications because of the ease, power and flexibility it offers in software development. Web applications producing output using XML/XSLT technology allows three types of classification options; classification at the source (XML classification), classification at the destination (HTML classification), and a new alternative: classification at the point of XSLT transformation. We have explored the third option for classifying web pages and showed that it is not only viable but also a preferable approach over the others as it takes advantages of both approaches. This technique is able to combine both the source and the destination document for better classification. More specifically, it is able utilize both structural data in XML and relevant data in HTML using the transformation rules in XSLT stylesheets. As a result a technique with a considerably higher classification rate is obtained.

We implemented a framework that incorporates the XSLT classification in a practical manner to classify web pages. In this framework different structural models and alternative classifiers can be combined to classify documents generated by XSLT Stylesheets.

Even though many e-business applications are using XSLT internally to generate and share XML/HTML documents, applications that rely on client side XSLT is rare. Although there are browsers with built-in XSLT processor, other types of clients such as cell phone, PDAs, TV sets do not have widespread XSLT support at present. This situation restricts applications to server side transformations in todays applications. With the availability of larger public datasets in the future, further experiments can be performed.

## References

[1] Engin Tozal, "Classification Using XSLT" MS Thesis, Comp. Eng.Fatih University, 2005.

[2] Atakan Kurt, Engin Tozal, "A Web Classification Framework Based on XSLT" ADWeb 2006 Lecture Notes in Computer Science (LNCS) 3842, pp. 86 – 96, 2006.

[3] S. Dumais, et al, "Inductive learning algorithms and representations for text categorization", 7th Int. Conf. on Information and knowledge management, pages 148--155. 1998.

[4] Thorsten Joachims, "Text categorization with support vector machines: learning with many relevant features", 10th European Conference on Machine Learning (ECML), 1998.

[5] V. N. Vapnik, The Nature of Statistical Learning Theory", Springer, 2nd edition, 1999.

[6] A. Basu, C. Watters, and M. Shepherd, "Support Vector Machines for Text Categorization", Proceedings of the 36th Annual Hawaii International Conference on System Sciences, 2003.

[7] Dunja Mladenic, "Turning Yahoo to Automatic Web-Page Classifier", European Conference on Artificial Intelligence, 1998

[8] F. Esposto, D. Malerba, L. D. Pace, and P. Leo. "A machine learning apporach to web mining", In Proc. of the 6th Congress of the Italian Association for Artificial Intelligence, 1999

[9] A. Sun and E. Lim and W. Ng, "Web classification using support vector machine", the 4th Int. Workshop on Web information and Data Management. ACM Press, 2002

[10] Arul Prakash Asirvatham, Kranthi Kumar Ravi, "Web Page Classification based on Document Structure", 2001

[11] H.-J. Oh, et al "A practical hypertext categorization method using links and incrementally available class information", the 23rd ACM Int. Conf on R & D in Information Retrieval 2000

[12] Soumen Chakrabarti and Byron E. Dom and Piotr Indyk, "Enhanced hypertext categorization using hyperlinks", Proceedings of the ACM SIGMOD, 1998

[13] Jeonghee Yi and Neel Sundaresan, "A classifier for semi-structured documents", Proceedings of the 6th ACM SIGKDD 2000.

[14] Ludovic Denoyer and Patrick Gallinari, "Bayesian network model for semi-structured document classification", Information Processing and Management, Volume 40, Issue 5, 2004.

[15] David D. Lewis, "Naive (Bayes) at Forty: The Independence Assumption in Information Retrieval" Lecture Notes in Computer Science; Vol. 1398, 1998.

[16] Irina Rish, "An empirical study of the naive Bayes classifier", IJCAI 2001 Workshop on Empirical Methods in Artificial Intelligence, 2001.

[17] Andrew McCallum and K. Nigam, "A comparision of event models for naive bayes text classification", AAAI-98 Workshop on Learning for Text Categorization, 1998.

# Machine Learning Models: Combining Evidence of Similarity for XML Schema Matching

Tran Hong-Minh and Dan Smith

School of Computing Sciences,
University Of East Anglia,
Norwich, UK
NR4 7TJ
{mtht, djs}@cmp.uea.ac.uk

**Abstract.** Matching schemas at an element level or structural level is generally categorized as either hybrid, which uses one algorithm, or composite, which combines evidence from several different matching algorithms for the final similarity measure. We present an approach for combining element-level evidence of similarity for matching XML schemas with a composite approach. By combining high recall algorithms in a composite system we reduce the number of real matches missed. By performing experiments on a number of machine learning models for combination of evidence in a composite approach and choosing the SMO for the high precision and recall, we increase the reliability of the final matching results. The precision is therefore enhanced (e.g., with data sets used by Cupid and suggested by the author of LSD, our precision is respectively 13.05% and 31.55% higher than COMA and Cupid on average).

## 1 Introduction

Comparing schemas to obtain matches is a major part of the schema matching process, information cooperation, data warehouse, e-commerce and query processing. In practice, schema matching is done manually with the help of graphical user interfaces in a labour-intensive process [4]. As the number of online information sources increases rapidly we need better ways of merging and summarizing information from multiple heterogeneous sources. Hence, improved schema matching algorithms and integration strategies are increasingly important.

Matching schema traditionally takes two internal schema representations (e.g., tree-like model or graph-like model) as an input and produces as the output a correspondence between the elements in the two input schemas. A matching problem is usually categorized as either hybrid or composite [10]. A common feature of all hybrid systems is that many criteria and schema properties (e.g. node label, data type, etc.) are exploited in a single algorithm. Hence, the order of comparing the criteria and properties is predefined and fixed in the algorithm, which is less flexible. On the contrary, in a composite system, numerous matchers are independently used. Each of them is described by one or more matching

R. Nayak and M.J. Zaki (Eds.): KDXD 2006, LNCS 3915, pp. 43–53, 2006.

algorithms and deals with different aspects of the schema. The final result is obtained by combining the results from each of the matching algorithm. Therefore the composite systems are more flexible.

In the other aspect, the aim of the matching activity is to classify each pair of elements of two representations into either the similarity category or the dissimilarity category. Hence, classification algorithms can be applied to matching problems. Studies in the fields of classification and data fusion (see e.g., [1, 7, 9]) have shown that a superior final result can be achieved by using a number of different algorithms and combing the results even when an individual algorithm performs poorly on its own, instead of using just a single algorithm. These studies suggest that the composite approach could give a better overall precision.

Motivated by above, we use a composite approach in our experiments, in order to obtain high precision matches while minimizing the number of missed matches. To match schemas in different aspects and obtain the results, we use six different algorithms, which are grouped into either syntax-driven techniques or semantic-driven techniques. The combination of those algorithms gives a complete view of similarity since each of them works on different aspects of schema element labels. To combine the evidence of similarity into the overall result we propose the use of machine learning models. The advantage of such models is that they avoid manually defining weights, thresholds or heuristics. The quality of which are largely determined by human expertise and domain knowledge.

We present machine learning methods to combine evidence of similarity computed by multiple algorithms into the overall matching determination. We present a set of experiments on machine learning models, by which we experimentally suggest a suitable model for high recall/precision. Our experimental results suggest that machine learning models in a composite system could significantly enhance the precision.

The rest of the paper is organized as follows. Related work is briefly presented in Section 2. An overview of the algorithms used in our element-level matchers and relationships between those matchers are given in Section 3. Section 4 discusses about methods for combining evidence and experimental results are provided in Section 5. Finally, we summarize our work in Section 6.

## 2 Related Work

Studies on XML Schema matching use either the hybrid or the composite approach. Most research has been based on the hybrid approach.

Cupid [2] is a hybrid system for schema matching at both element and structure levels. In the Cupid system, the schema matching is produced by deducing a match from computed similarity coefficients between elements of the two schemas. Cupid combines matching information by firstly using data type and synonym information from tokenized and categorized labels, and then following a bottom-up structure-based matching which exploits information from the node's immediate parent, to get more precise matches. Two elements are similar if their leaf sets are similar. The similarity of the leaves is increased if they have ancestors that are similar. The final similarity coefficients, $\text{WordSim}_{\text{Cupid}}$, is the weighted combination of the similarity at the structure-level matching, structureSim, and the label meaning similarity at the element-level matching, labelSim. The $\text{weight}_{\text{struct}}$ defines the contribution of each level matching to the final similarity degree:

$$\text{WordSim}_{\text{Cupid}} = \text{weight}_{\text{struct}} \times \text{structureSim} + (1 - \text{weight}_{\text{struct}}) \times \text{labelSim}. \quad (1)$$

XClust [8] is another hybrid system for both element level and structure-level matching. It proposes a matching function with thresholds to score DTDs for deducing matches. At the element level, Unlike Cupid, which uses a built-in dictionary and a number of different component properties, in XClust the label similarity solely results from WordNet dictionary [5, 11]. The use of WordNet is limited into exploiting synonyms in XClust system. The degree of structural similarity is determined by tree-edit distance algorithm. In XClust, the label meaning—SemanticSim, the number of common children at the leaf level—LeafContextSim, and the number of common direct children ImmediateDescSim are counted for the final similarity—ElementSim—of two elements $e_1$ and $e_2$:

$$\begin{aligned} \text{ElementSim}(e_1, e_2) = \quad & \alpha \times \text{SemanticSim}(e_1, e_2) \\ & + \beta \times \text{LeafContextSim}(e_1, e_2) \\ & + \gamma \times \text{ImmediateDescSim}(e_1, e_2) \end{aligned} \quad (2)$$

where $\alpha + \beta + \gamma = 1$ and $\alpha$, $\beta$ and $\gamma$ are weights of SemanticSim, LeafContextSim and ImmediateDescSim, respectively.

Unlike the Cupid and XClust approaches, the LSD system [4] is a composite system using a multi-strategy learning approach for element-based one-to-one matching at the leaf level of tree-like schemas. Each engine learns well certain kinds of patterns and then the predictions of the learners are combined by using meta-learners.

Similar to LSD, COMA [3] is a composite generic matching system but different from LSD, it consists of multi-matchers which do not use machine learning techniques. To combine results computed from various matchers, it supports different rule-based methods, for example, taking the average or the maximum value of results. It allows the user to interact the match process by feedbacks and to choose a combination of results from its extensible library.

However, the composite approach is not so widely studied as the hybrid approach. In the composite approach, current matchers and combinations are categorized into two types. In the first type, both matchers and the combination have been in the context of machine learning. In the second type, matchers are not confined to machine learning but the combination method is simple rule-based. Our matchers are also not restricted to machine learning. To combine similarity evidence we are not restricted into rule-based, we use a formal learning model, which is more generic and good-result promising.

## 3 Element-Level Matchers Overview

At the atomic level, we use the composite approach which includes six different matchers. Results produced by matchers are combined in a formal model to determine if a pair of nodes makes a match. The element level matching consists of three macro steps:

Step 1: **Pre-processing**: The objective is to make all schemas comparable. Many semantically similar component labels contain abbreviations and acronyms that make them syntactically different due to different style of encoding a XML Schema. Furthermore, some of our comparisons for a matching candidate take into account each word of a label with every word in the other label instead of the whole labels. Redundant words in a label thus could reduce the similarity degree of the two labels. To avoid those obstacles, we tokenize labels into words, expand abbreviations and acronyms into the original forms and build a minimum set of words presenting the same concepts denoted by before-pre-processed labels.
Step 2: **Computing similarity**: For each pair of nodes in two tree-like structures, we compute independently the similarity of each pair by using six different matchers.
Step 3: **Determining final similarity:** we use a learning model to determine whether or not the pair is similar based on six results computed by matchers. Step 3 is discussed in details in Section 4.

The meaning of a label can be investigated in three different aspects: its lexical view, its semantic meaning and phonetic view. In Step 2, each matcher covers a certain aspect of labels then the combination of results could give a complete and appropriate prediction of the similarity of two labels. Therefore, in Step 2, we use six different matchers covering all such aspects to determine the similarity of a pair of labels. The matchers implementing string-based algorithms [6](e.g., edit distance, approximate string and q-gram) explore lexical view of labels. Those implementing sense-based algorithms [11] (e.g., WordNet sense analysis) directly cover the semantic meaning of concepts denoted by labels. Those implementing phonetic-based algorithms [6] (e.g., soundex and metaphone) exploit the phonetics of labels.

## 4 Combining Similarity Evidences

Sub-results have to be combined together for obtaining a final result. A popular way for the combination is to establish a formula with weights and thresholds. Each result is experimentally assigned a weight on the basis of researcher's experiences about the domain and/or the importance of the matcher to the whole system. The thresholds are also defined by a similar method. To cover some exceptions, heuristic rules are used. However, it is easy to see that the formula with weights, thresholds and heuristic rules are not flexible and could not embrace all possible cases.

In the case of using a single matcher and combination from multiple matching criteria, such as Cupid or XClust, in formula (1) of Cupid and in formula (2) of XClust, the weights $weight_{struct}$ for Cupid and $\alpha$, $\beta$ and $\gamma$ for XClust are all predefined to reflect the fixed contributions of each matching phase for whatever matching cases are. They therefore are not tolerant to the variations in matching cases.

In the case of using multiple matchers, we also suggest that the matching function and thresholds cannot deliver a good result. Various results obtained by independent matchers could have conflict or agreement, which are varied and difficult to describe in a formula. For example, {Contact, Person, Primary} and {Primary, Contact, Person} have sense-based similarity degree of 0.3, which indicates that they are not similar. It is correct as they have different meanings. Whereas, they get the value of 0.947 returned by the approximate string matching algorithm, which indicates that they are very similar. This is an incorrect result. Therefore, in this case, the similarity determined by the sense-based algorithm is against the result of the approximate string matching algorithm and it should be dominant the final result as it is correct. However, for example, {Postcode} and {Postal, Code} which have the value of 0.8 and 0.75 respectively computed by the sense-based algorithm and the approximate string matching algorithm. In this case, those values have agreement.

Therefore, machine learning models are intuitively more suitable to determine the final result on the basis of sub-similarity degrees. Especially, as most learning models work well for the problems of binary classification, they are more applicable for our matching determination, which is either similarity or dissimilarity rather than a degree.

Regarding using machine learning models for fusion, evidence sets which are good candidates for an effective combination are very important for a good final result. Therefore matchers which produce good evidence sets for the fusion are crucial. It raises a problem of defining characteristics of good matchers. Based on [9, 12], we define three characteristics of good matchers as follows:

(a) At least one matcher produces high accuracy results,
(b) High overlap of real matches (true positive),
(c) Low overlap of real non-matches (true negative).

The above criteria (b) and (c) are described by the below formulas:

$$R = \frac{n \times R_{\text{common}}}{\sum_{i=1}^{n} R_i}, \quad N = \frac{n \times N_{\text{common}}}{\sum_{i=1}^{n} N_i}, \tag{3}$$

where $R$ and $N$ are the overlap ratios of real matches and real non-matches, respectively. $R_i$ and $N_i$ are the real matches and the real non matches respectively computed by the $i$-th algorithm in the system. $R_{\text{common}}$ and $N_{\text{common}}$ are common portions of real matches and of non-matches of all result sets, respectively.

## 5 Experiments

Three sets of experiments are done for three purposes. We compare various machine learning models for final determination from multiple results in the first experimental set. Those experiments give an idea about a suitable formal model for the fusion of evidence. We analyze the combination of six algorithms proposed in Section 4 on the basis of characteristics in the second experimental set. The discussion shows how suitable the algorithms are to be good candidates for fusion. We do some comparisons between our approach with two existing ones, e.g., Cupid and COMA in the last experimental set.

We use a 10-fold cross-validation method for all our experiments. To evaluate the matching quality we use the following indicators: recall, precision, and F-measure. To evaluate the time efficiency we compare computational time spent for building the data model for each learning model. All experiments are performed on a Intel Centrino 1.7GHz with 768MB RAM and WindowsXP operation system. All implementations are written in Java without using any compiler optimization.

### 5.1 Comparison of Models for Combining Evidences

In the first set of experiments, we use the same quite large XML Schema data sets with 739 components (includes elements and attributes), published by FGDC [1] (Federal Geographic Data Committee) and NBII [2] (National Biological Information Infrastructure). For each pair of components we measure sub-similarity degrees by using matching algorithms and then combine them to determine if a pair is a match by using various learning models. The experimental results produced by each model are then compared statistically to choose the most accurate model.

**Table 1.** Computational time (in second) of 18 learning models

| Models | time | Models | time | Models | time |
|---|---|---|---|---|---|
| Bays.Net. | 5.49 | Nave | 2.03 | Compl.Naive | 0.19 |
| NaiveUpdate. | 1.86 | Logistic | 11.06 | Bagging | 70.49 |
| RBFNet. | 38.87 | ThresholdSelector | 19.29 | SimpLogistic | 278.14 |
| FilterClass. | 6.45 | SMO | 5.51 | LogitBoost | 53.21 |
| VotedPercpt. | 28.41 | MultiClassClass. | 9.82 | AdaBoostM1 | 27.56 |
| OrdClassClass. | 22.59 | AttrSelClass. | 6.63 | RacedIncrLogBoost. | 12.23 |

All 18 learning algorithms compared in this set of experiments mostly inferentially explore the undetermined relationships among variables. The algorithms are classified into three categories: Bayes model, function learning model and

[1] http://www.fgdc.gov/. FGDC is responsible for metadata about geographic data

[2] http://www.nbii.gov/. NBII offers metadata standard for images related to nature and the environment.

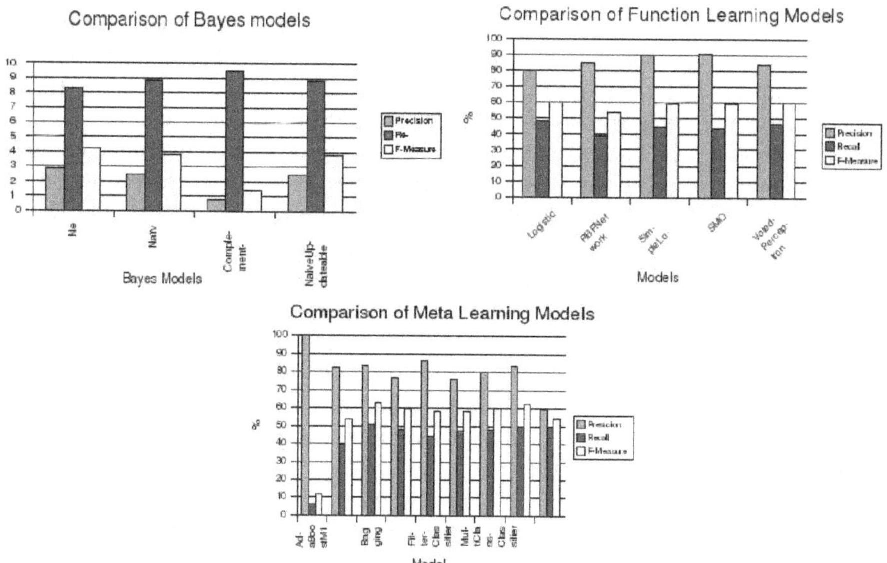


**Fig. 1.** Precision, Recall and F-Measure of (a) Bayes models (b) Function learning models (c) Meta learning models

meta learning model. We use evidence values computed by matchers as variables for the learning models to investigate the relationships. The inferences are then used for determining if two labels are matched.

In the Bayes category, we compare Baysian network, Naïve Bayes, Complement Naïve Bayes and Updateable Naïve models. The function learning category has Logistic, RBF Network, Simple Logistic, SMO and Voted Perceptron models. In the meta learning category, we have Ada BoostM1, Attribute Selected Classifier, Bagging, Threshold Selector, Filter Classifier, LogitBoost, Multi-Class Classifier, Ordinal Class Classifier and Raced Incremental Logic Boost models.

Figure 1(a), (b) and (c) display precision/recall and F-measure, which indicate the quality of algorithms. Table 1 shows computational time for building model of Bayes models, function learning models and meta learning models, respectively.

In Figure 1, among the Bayes models, the Bayesian Network has the highest precision and its recall is not so different to other Bayes models. In general, among models in three categories, Bayes models give the highest recall (88.35% in average), which means that the number of real matches missed is the smallest. The precision is also the lowest (21.38% in average), which means they deliver the maximum number of unreal matches. Hence, matches obtained by such models are not reliable. However, as having the highest recall, Bayes models are more applicable for recognizing and eliminating non-matches.

Models in the function learning category and the meta learning category give higher precision (average 85.5%, 78.6% ) but lower recall (average 44.5%, 47.3% )

than Bayes models. Hence, function learning models and meta learning models are more reliable for the matching problem than Bayes models are. In the meta learning category, the Filter Classifier model gives the highest precision. In the function learning category, both SMO model and Simple Logistic model achieve the highest precision. They also deliver the highest precision on overall, compared with all models from the three categories. SMO produces 90.4% precision, 44.1% recall and 59.2% F-measure. Simple Logistic model produces 89.9%, 44.4% and 59.4% in precision, recall and F-measure, respectively.

Regarding computational time, from our observation on experimental results, in general, the higher precision/recall and F-measure a model achieves the more expensive in computational time could be. Bayes models, which have the lowest reliability, have the smallest computational time (2.39s in average). The function learning models, which have the highest reliability, have the highest computational time (72.40s in average). Furthermore, in the Bayes category, the Bayesian Network, which has the highest precision, also has the highest computational time. Similarly, models giving high precision (such as Bagging, Simple Logistics) also have high computational time. However, SMO and Filter Classifier which are among the best models for high precisions have small computational time.

### 5.2 Analysis of Combining Evidences

Section 5.1 shows that Simple Logistic, SMO and Filter Classifier learning models give best precision/recall and computational time. Therefore, in this experimental set, we use them for the analysis of our fusion of evidence returned by matching algorithms. The quality of combining algorithms are analyzed on the basis of three characteristics proposed in Section 4.

We measure precision/recall and F-measure for each algorithm. Table 2 shows precision/recall and F-measure of each algorithm with the default threshold value of 0.5. We observe that Metaphone algorithm gives the highest precision and WordNet algorithm gives the highest recall. Thus, the first characteristic is satisfied.

Furthermore, before fusion, the average precision of all algorithms is 45.74% but the precision after fusion by using SMO model is 91.9%. Obviously, using multiple algorithms and combining them significantly improve the precision. However, the recall before fusion is 70.76% but it is 59.6% after using SMO for combining results.

**Table 2.** Precision-Recall-F-measure (in %) of each algorithm without using learning models

| | EditDist. | WordNet | Apprx.Str | Soundex | Metaphone | Q-Gram |
|---|---|---|---|---|---|---|
| Precision | 41.05 | 12.27 | 44.55 | 23.78 | 96.77 | 56.00 |
| Recall | 68.42 | 82.46 | 78.95 | 68.42 | 52.63 | 73.68 |
| F-measure | 51.32 | 21.36 | 56.95 | 35.29 | 68.18 | 63.64 |

**Table 3.** Overlap of real matches and real non-matches in fusion of algorithms (in %)

| Algo.1 | Algo. 2 | Match | Non-match | Algo.1 | Algo. 2 | Match | Non-match |
|---|---|---|---|---|---|---|---|
| EditDist. | WordNet | 78.07 | 21.05 | WordNet | Metaphone | 77.92 | 24.39 |
| EditDist. | Approx. | 92.86 | 28.62 | WordNet | Q-Gram | 78.65 | 20.67 |
| EditDist. | Soundex | 92.31 | 28.62 | Approx. | Soundex | 90.48 | 26.48 |
| EditDist. | Metaphone | 84.06 | 35.33 | Approx. | Metaphone | 80.00 | 32.51 |
| EditDist. | Q-Gram | 91.36 | 30.05 | Approx. | Q-Gram | 96.55 | 28.02 |
| WordNet | Approx. | 82.61 | 19.72 | Soundex | Metaphone | 86.96 | 33.33 |
| WordNet | Soundex | 81.40 | 19.84 | Soundex | Q-Gram | 88.89 | 27.93 |
| All algorithms | | 71.90 | 59.29 | Metaphone | Q-Gram | 83.33 | 34.50 |

In overall, by using multiple matchers and evidence fusion we obtain more efficiency in matching, since the F-measure is higher than using only one matchers. The F-measures after using fusion are 59.2% for the SMO model and 59.4% for the Simple Logistic model, which are higher than the average F-measure of individual algorithms (49.46%).

We use formula (3) for analysing the the final characteristics raised in Section 4. We build the overlap ratios for real matches $R$ and for real non-matches $N$ for all possible pairs of algorithms(in Table 3). Most of our couples produce high match-overlap and small non-match-overlap ratios, which are satisfied the characteristics in Section 4. Therefore, our set of algorithms is good candidates for fusion of evidence.

### 5.3 Comparison with Cupid and COMA

We perform two experiments to compare our approach with COMA and Cupid. In the first experiment we use CIDX Schema and EXCEL schema, which were used by Cupid. The first data set has 68 components (elements and attributes). In the second experiment we use schemas about courses in Cornell University and Washington University. The second data set has 73 components. Suggested by above experimental results, we use SMO learning model for determining if two components are a match.

The Figure 2 (a) and (b) show that our approach is better, especially in the first experiment. In the second experiment, our precision is nearly equal to the COMA precision and better than Cupid one, but our recall is not. Table 4 also shows that our approach is more efficient in time.

**Table 4.** Comparison of computational time for building data models(in second), using CIDX–EXCEL data set and Course (of Cornell Uni. and Washington Uni.) data set

| | CIDX-EXCEL | | | Course | | |
|---|---|---|---|---|---|---|
| | Cupid | COMA | Ours | Cupid | COMA | Ours |
| Computational time | 0.60 | 1.70 | 0.45 | 10.00 | 3.00 | 1.35 |

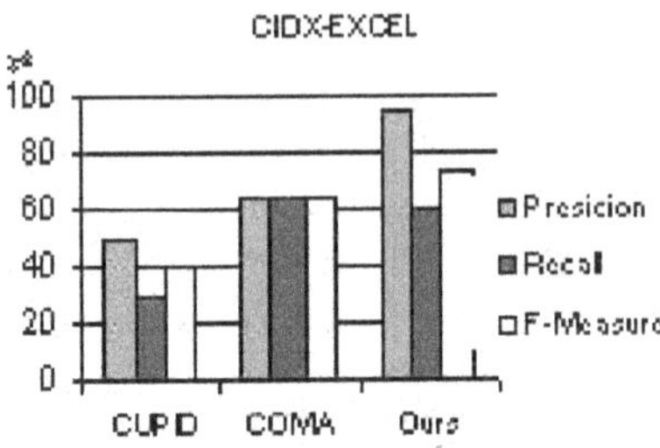


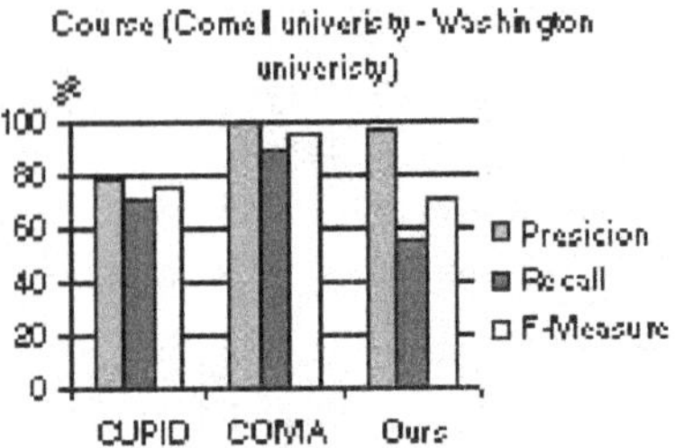


**Fig. 2.** Comparison on (a) CIDX–EXCEL data set (b) Course (of Cornell Uni. and Washington Uni.) data set

# 6 Conclusion

In this paper, we present a composite approach with multi-matchers for element-level XML schema matching. The approach takes into account both syntactic and semantic matching. It uses six different matching algorithm for dealing all aspects of schema element labels. We analyze to show that relations between results are difficult to model into a formula. We propose a machine learning method to determine the final results on the basis of results computed by matchers and we achieve a better precision than Cupid and COMA did (e.g., with data sets used by Cupid and suggested by the author of LSD, our precision is respectively 13.05% and 31.55% higher than COMA and Cupid in average.). By carrying out experiments on 18 machine learning models, we see that the SMO learning model gives the best performance.

To improve the recall, we plan to exploit data type and instance for improving matching results at both the element level and the structural level. Besides improving matching algorithms themselves, we plan further work on using multiple learning models for a higher accuracy, since we recognize from our experimental results that some models can produce high recall but low precision and vice versa.

# References

1. B. T. Bartell, G. W. Cottrell, and R. K. Belew. Automatic combination of multiple ranked retrieval systems. In *17th annual international ACM SIGIR conference on Research and development in information retrieval*, pages 173–181, 1994.
2. P. A. Bernstein, J. Madhavan, and E. Rahm. Generic schema matching with cupid. In *27th VLDB*, volume 10, pages 49–58, 2001.
3. H.-H. Do and E. Rahm. Coma - a system for flexible combination of schema matching approaches. In *VLDB*, pages 610–621, 2002.
4. A. Doan, P. Domingos, and A. Y. Levy. Learning source description for data integration. In *WebDB*, pages 81–86, 2000.
5. C. Fellbaum, editor. *WordNet: An Electronic Lexical Database.* MIT, Cambridge, MA, 1998.

6. P. A. V. Hall and G. R. Dowling. Approximate string matching. *ACM Comput. Surv.*, 12(4):381–402, 1980.
7. J. H. Lee. Analyses of multiple evidence combination. In *20th annual international ACM SIGIR conference on Research and development in information retrieval*, pages 267–276, 1997.
8. M. L. Lee, L. H. Yang, and W. Hsu. Xml schemas: integration and translation: Xclust: clustering xml schemas for effective integration. In *CIKM*, pages 292–299, 2002.
9. M. C. McCabe, A. Chowdhury, D. Grossman, and O. Frieder. System fusion for improving performance in information retrieval systems. In *International Conference on Information Technology: Coding and Computing (ITCC '01)*, pages 639–644, 2001.
10. E. Rahm and P. A. Bernstein. A survey of approaches to automatic schema matching. *The VLDB Journal*, 10(4):334–350, 2001.
11. N. Seco, T. Veale, and J. Hayes. An intrinsic information content metric for semantic similarity in wordnet. In *ECAI*, pages 1089–1090, 2004.
12. C. C. Vogt and G. W. Cottrell. Predicting the performance of linearly combined ir systems. In *21st annual international ACM SIGIR conference on Research and development in information retrieval*, pages 190–196, 1998.

# Information Retrieval from Distributed Semistructured Documents Using Metadata Interface

Guija Choe[1], Young-Kwang Nam[1],
Joseph Goguen[2], and Guilian Wang[2]

[1]Department of Computer Science, Yonsei University, Wonju, Korea
gjchoe@hosu.yonsei.ac.kr, yknam@dragon.yonsei.ac.kr
[2]Department of Computer Science and Engineering, UCSD, La Jolla, CA 92093
{goguen, guilian}@cs.ucsd.edu

**Abstract.** We describe a method for retrieving information from distributed heterogeneous semistructured documents, and its implementation in the metadata interface DDXMI (Distributed Document XML Metadata Interface). The system generates local queries appropriate for local schemas from a user query over the global schema and shows the result of the generated queries. The three components are designed to generate the local queries: mappings between global schema and local schemas (extracted from local documents if not given), path substitution, and node identification for resolving the heterogeneity among nodes with the same label that often exist in semistructured data. The system uses Quilt as its XML query language. An experiment is reported over three local semistructured documents: 'thesis', 'reports', and 'journal' documents with 'article' global schema. The prototype was developed under Windows system with Java and JavaCC.

## 1 Introduction

There is much research on integrating distributed heterogeneous data with explicit schemas, which are called structured data. Besides expensive data warehousing, a major focus is virtual integration, i.e., developing portals that allow uniform querying through a global schema to distributed heterogeneous data [12], [14], [18], [21], [25]. A query over the global schema is usually resolved and answered by consulting mappings between the global and local schemas. Semistructured data models emerged as a result of the efforts to extend database management techniques to data with the irregular, unknown, and frequently changing structures that are becoming more and more common as the Internet grows [1], [2], [13]. However, for semistructured data, structural information is not given explicitly, and usually data are created without any restriction on structure, so that it is much more difficult to develop such data processing system. Because the semistructured data have no specific rules or enforcement of the structure, it often happens that elements with the same tag have different

R. Nayak and M.J. Zaki (Eds.): KDXD 2006, LNCS 3915, pp. 54–63, 2006.

structures and contain different information so that a single element in the global schema may correspond to several elements with the same tag and even the same path in an extracted local schema with different mapping types, i.e., 1:1, 1:N, and N:1 mappings.

We designed a system to address this problem and implemented it in a research system for generating local queries over distributed semistructured documents through a metadata interface and the queries are executed on its own local site. It handles semistructured data with additional functionality to extract schemas for XML documents without explicit schema information, as identifying different types of elements with the same tag in our query processing system. The proposed system architecture is shown in Fig. 1. Queries over the global schema are processed based on the mapping information stored in a structured document called DDXMI (for Distributed Documents XML Metadata Interface), which works as an integrated view over all relevant local schemas. The DDXMI file contains the mapping information and functions to be applied to each local document, along with some identification information such as author, date, comments, etc. The system prototype has two parts: the DDXMI Generator for mapping the global schema with local schemas and producing a DDXMI file, and the Query Generator for generating the local queries and answering queries. Our tool parses a document schema or the document itself if its schema is unknown to get the structure of the document, and then generates a dynamic path tree, which can be folded and unfolded by clicking. The mapping is specified by assigning indices through clicking involved nodes in the path trees in a GUI, which link local elements to corresponding global elements and to the names of conversion functions. These functions can be built-in or user-defined in Quilt [7], which is our XML query language. The DDXMI document is then generated by collecting over index numbers, which are internal to the system. User queries are

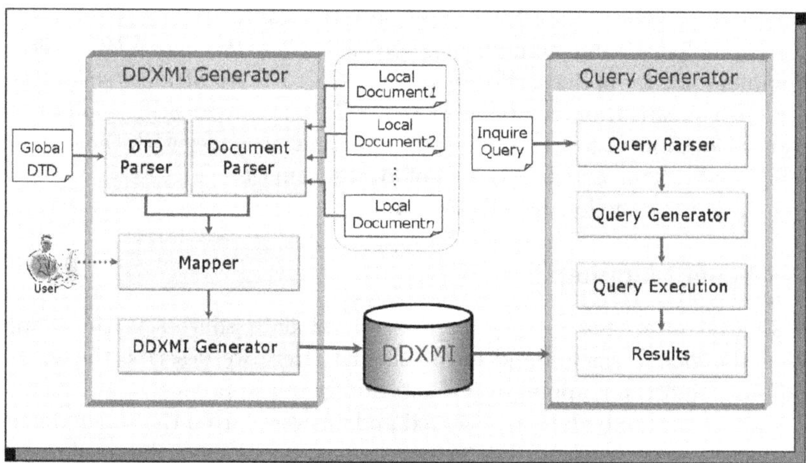


**Fig. 1.** The structure of the proposed system

rewritten into appropriate queries for each relevant local document according to the mapping information in the DDXMI document and node identification information; finally each local query is processed by Kweelt engine for Quilt.

## 2 The Related Work

To facilitate formulation, decomposition and optimization of queries for semistructured data, schema extraction or type inference have been studied by using machine learning methods and heuristics [19], [20], [22]. Unfortunately, the accuracy goes down as the extracted schema size decreases.

Schema mapping is a critical step for data integration and many other important database applications. An extensive review of techniques and tools for automatic schema matching up to 2001 is given in [23]. Traditional approaches such as instance-based LSD [8] and GLUE [9] and schema-based Cupid [15], SF [16], Rondo [17] and Coma [10], and the holistic approach MGS [11] only help find 1-to-1 matches, and have great difficulty with matches that involve conditions or conversion functions, and cannot discover n-to-m matches for n>1 or m>1, automatically. Some tools such as COMAP, Clio, SCIA find complex matches based on user input [4], [5], [6], [9], or ontologies [24]. However, it is very difficult for these tools to deal with extremely complex mappings where schema nodes have the same label but different types (these often exist in semistructured data).

Rewriting queries using views has been studied extensively for structured data [12], [18] and for semistructured data [3]. But those researches and techniques all targeted at restricted formats of views.

## 3 The Three Query Processing Components

Our method for generating a set of appropriate local queries $Q_{out}$ from a global query $Q_{in}$ includes three components, *M(LSS, GS)*, *PS*, and *NIP*, where *M* is a component for mapping a global schema *GS* to a set of local schemas *LSS*, *PS* is a path substitution component, and *NIP* is a node identification predicate generation component for resolving the heterogeneous nodes with the same label, *GS* is a global schema and *LSS* is a set of local schema $LS_1$ , . . . , $LS_j$. We describe each of these in the following sub-sections.

### 3.1 Schema Mapping

The essential part of a system for distributed data sources is the mappings between the global schema and local schemas. Here we describe the semantics of mappings and the mapping representation in our approach.

Assuming that only data are queried and answered from the single document for each site and there are no JOINs among local documents, the total mappings *M(LSS, GS)* are the union of the mappings of the global schema to each of the local schemas, *M(LS, GS)*. Let *G* and *L* be the set of nodes in the path tree *GT* of

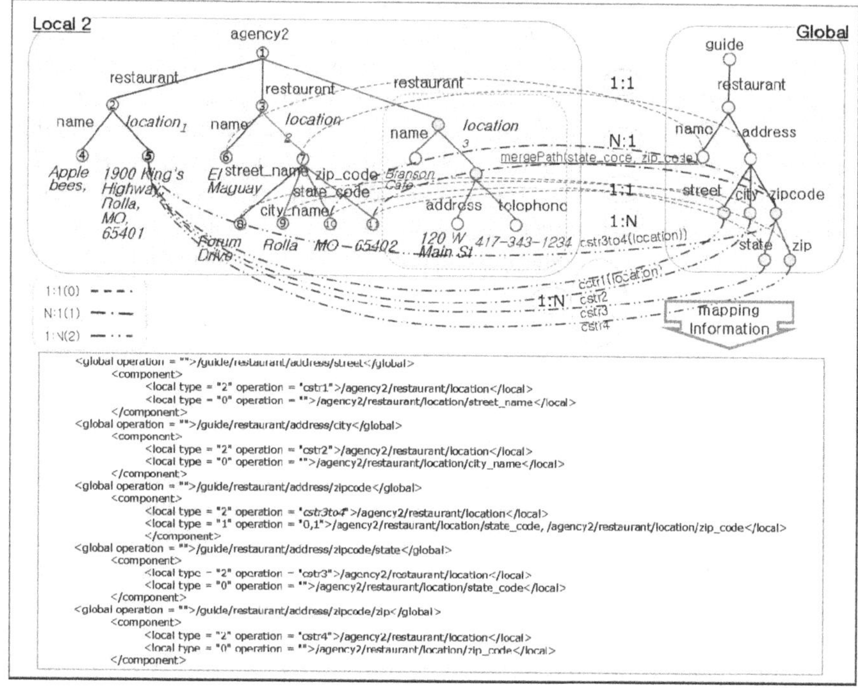


**Fig. 2.** Mappings between Global and Local nodes

the global schema *GS* and the set of nodes in the path tree *LT* of a local schema *LS* respectively, and let *PG* and *PL* be the power set of *G* and *L* respectively. A node $o_i$ in a path tree is an object $(ol_i,\ ov_i)$ which consists of the node label $ol_i$ and the node value $ov_i$. In Fig. 2, the node number *5* has the node label 'location' and the value '1900 King's Highway, Rolla, MO, 65401'. In *GT* and *LT*, several nodes may have the same labels, so we differentiate them by putting the subscript in the label when necessary, such as 'location$_1$' and 'location$_2$'. The mappings *M(LS,GS)* between the global schema *GS* and the local schema *LS* contain mapping elements in the format of $(l,\ g) \in PL \times PG$, where $g = (gn_1,\ gn_2,\ \ldots,\ gn_m) \in PG$ and $gn_i \in G$ for $i = 1\ to\ m$ and $l = (ln_1,\ ln_2,\ \ldots,\ ln_n) \in PL$ and $ln_i \in L$ for $i = 1\ to\ n$. We group the mapping elements according to their mapping types as follows:

$$M(LS,GS) = M_{11}(LS,GS) \cup M_{1N}(LS,GS) \cup M_{N1}(LS,GS) \text{ where}$$

i) $M_{11}(LS,\ GS)$ is the set of one-to-one mapping elements $m_{11} = (l,\ g)$, where $g \in PG$ and $l \in PL$ are both the singleton.
ii) $M_{1N}(LS,\ GS)$ is the set of one-to-many mapping elements $m_{1N} = (l,\ g)$, where $g = (gn_1,\ gn_2,\ \ldots,\ gn_m) \in PG$, $m>1$, $l \in PL$ is a singleton.
iii) $M_{N1}(LS,\ GS)$ is the set of many-to-one mapping elements $m_{N1} = (l,\ g)$, where $g \in PG$ is a singleton and $l$ is $(ln_1,\ ln_2,\ \ldots,\ ln_n) \in PL$, $n>1$.

In Fig. 2, 'guide' and 'agency2' are the names of the global and local schemas respectively, $M_{11}$ *(agency2, guide)* is ((state-code, state), (zip-code, zip)), $M_{1N}$ *(agency2, guide)* is ((location, (street, city, zipcode)), (location, (street, city, state, zip))), and $M_{N1}$ *(agency2, guide)* is ((state-code, zip-code), zipcode).

For $m = (l, g)$ where $l$ is not a singleton, functions are required for combining the content of multiple elements of $l$ into an instance of $g$. Even for $m = (l, g)$ where $l$ is a singleton, conversion functions are often required for transforming the content of $l$ into an instance of $g$. We call both combining and conversion functions as transformation functions. Therefore, the transformation $T_m$ over $m = (l, g) \in M(LS, GS)$ is $T_m{:}l \rightarrow g$ where $T_m$ is a vector of functions applied to the values of objects in $l$ in order to get the appropriate values for objects in $g$, i.e., $T_m(l) = g$, where $|T_m| = |g|$.

### 3.2 Path Substitution for Generating Local Queries

Quilt is used as the XML query language in our prototype. A typical Quilt query usually consists of FOR, LET, WHERE and RETURN clauses. FOR clauses are used to bind variables to nodes. In order to identify some specific nodes, more condition may be given inside of '[ ]' predicate. Therefore, path substitution in FOR clauses and WHERE clauses vary according to the mapping kind. In case of N:1 mapping, one global path is mapped by N local paths in a single local document, multiple variables may be introduced for those N nodes, or the parents of the N local nodes are bound and give conditions in predicates. When comparison of node values is involved, relevant transformation functions have to be combined with the paths during path substitution. The primary work for the local query generation from global queries is to replace paths in the global query by the corresponding paths appropriate to the local documents.

For example, in Fig. 2, *PS*(address/zipcode) = (location/zip-code, location/ state-code) since the global element 'address' corresponds to the local element 'location' and the global element 'zipcode' maps to (state-code, zip-code) for many-to-one mapping along the 'location' path, hence *PS*(address/zipcode) = $T_m$(address/zip-code) = *mergepath*(location/state-code, location/zip-code). *PS* (address/street) = (location) along the 'location' path since there is no mapping for 'address', and 'street' maps to 'location', hence *PS*(address/street) = $T_m$(address/street) = $cstr_1$(location).

### 3.3 Resolving the Heterogeneity of Nodes in Local Documents

Recall that the primary difference between structured and semistructured data is that a semistructured document may have several nodes with the same name but different structures. In this case, the nodes with the same label but different structures may map to multiple global nodes in different ways; some may even be mapped and some not, so a condition statement indicating that some unmapped nodes should not participate in the possible candidate answer is needed in the output local query.

For example, in Fig. 2, consider a global query given as Query1 and assume 'address' in 'guide' is only mapped to the nodes '$location_1$' and '$location_2$' node and not to '$location_3$'. The '$location_3$' node is not relevant to this query. Thus, the local query generator checks whether there are irrelevant nodes to the global query in the local path tree. If so, then such nodes must be explicitly screened by using path filtering predicates.

```
[ Query1 : A global query for 'guide' schema ]
  FOR $addr IN document("guide.xml")//address
  WHERE $addr/zipcode[CONTAINS(.,"MO")]
  RETURN $addr
```

Let $l_i$ and $l_j$ be two nodes with the same label but different structures in a local path tree. If $l_i$ and $l_j$ are mapped to the same global node, then $l_i$ and $l_j$ are called homogeneous, otherwise they are said to conflict. All the nodes sharing the same label with $l_i$ and mapped to the same global node are represented as a set, *homo*($l_i$), while the set of nodes conflicting $l_i$ is *conflict*($l_i$). In Fig. 2, *homo*($location_1$) = {$location_1$, $location_2$}, *conflict*($location_1$) = {$location_3$}, and *conflict*($location_2$) = {$location_3$} since 'address' is mapped to '$location_1$' and '$location_2$' but not to '$location_3$'. In Query1, the CONTAINS(.,"MO") predicate is applied to the '$location_1$' and '$location_2$' nodes, but not to '$location_3$' since 'address' maps to only '$location_1$' and '$location_2$'. To select the homogeneous elements '$location_1$' and '$location_2$', some specific conditions need to be specified.

Let $ln_i \in l$ and $L_{hc}(ln_i)$ be the set of nodes having the same label, but different structure so different index numbers, hence $L_{hc}(ln_i) = homo(ln_i) \cup conflict(ln_i)$. For any element $ln_i$, *childpaths*($ln_i$) is defined as the set of paths from $ln_i$'s direct children to leaf nodes. The super child path set $SCP(ln_i)$ of $ln_i$ is defined as the set of all child paths for all elements of $L_{hc}(ln_i)$, i.e., $SCP(ln_i) = U_{i=1}^{k}$ *childpaths*($h_i$), where $h_i \in L_{hc}(ln_i)$, $k = |L_{hc}(ln_i)|$. We use *childpaths*($ln_i$) and $SCP(ln_i)$ to formulate predicates to specify only node $ln_i$ while excluding any other nodes sharing the same label with $ln_i$. The predicate (($p_1$ AND, ..., AND $p_i$) AND (NOT($q_1$) AND NOT($q_2$), ..., AND NOT($q_j$))) for $p_i \in$ *childpaths*($ln_i$) and $q_i \in$ ($SCP(ln_i)$ - *childpaths*($ln_i$)) means that $ln_i$ has the child paths $p_1, \ldots, p_i$ and should not have the child paths $q_1, \ldots, q_j$.

# 4 System Implementation and Execution Examples

## 4.1 Mapping Representation and Path Substitution

The mapping information for the global schema and local schemas is stored in a structured XML document, a DDXMI file. The structure of DDXMI is specified in DDXMI's DTD, shown in Fig. 3. The elements in the global schema are called global elements, while the corresponding elements in the local documents are called local elements. When the query generator reaches a global element name in a global query, if its corresponding local element is not null, then the paths in the query are replaced by the paths to the local elements to get local queries.

```
<!ELEMENT DDXMI (DDXMI.header, DDXMI.isequivalent, documentspec)>
<!ELEMENT DDXMI.header (documentation,version,date,authorization)>
<!ELEMENT documentation (#PCDATA)>
<!ELEMENT version (#PCDATA)>
<!ELEMENT date (#PCDATA)>
<!ELEMENT authorization (#PCDATA)>
<!ELEMENT DDXMI.isequivalent (global, component*)*>
<!ELEMENT global (#PCDATA)>
<!ELEMENT component (local*)>
<!ELEMENT local (#PCDATA)>
<!ATTLIST global operation CDATA #IMPLIED>
<!ATTLIST local
    type CDATA #REQUIRED
    operation CDATA #IMPLIED
>
```

**Fig. 3.** The DDXMI's DTD

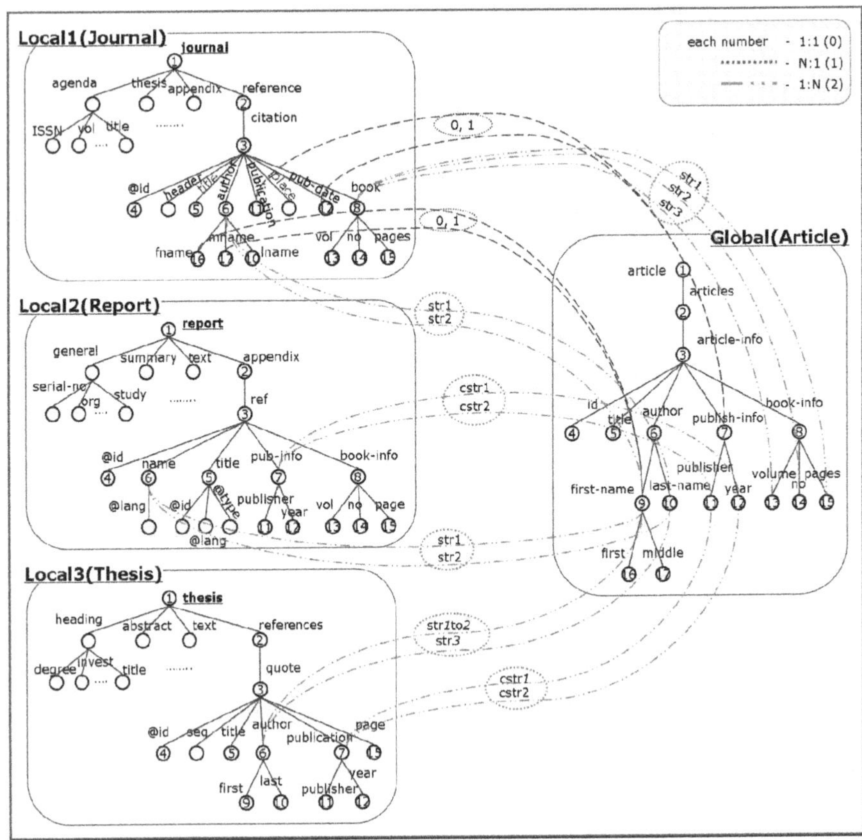


**Fig. 4.** A portion of the mapping information for 'article' global schema and 3 local documents

<table>
<tr><td colspan="2">Global Query (Article)</td><td><result><br>FOR $gs IN document("Article.xml")//author[//first-name]<br>WHERE $gs[CONTAINS(., "M")]<br>RETURN $gs<br></result></td></tr>
<tr><td rowspan="3">Generated Query</td><td>Local1 (Journal)</td><td>import mergePath as UDF merge;<br>import split as USER_split;<br>FUNCTION str1($str)<br>{ split(" ", "1", $str) }<br><result><br>(FOR $gs IN document("Journal.xml") //author[/fname AND /lname AND /mname][mergePath(/fname, /mname)]<br>WHERE mergePath($gs/fname, $gs/mname) [ CONTAINS (. , "M") ]<br>RETURN $gs) |<br>(FOR $gs IN document("Journal.xml") //author[NOT(/fname) AND NOT(/lname) AND NOT(/mname)][str1(.)]<br>WHERE str1($gs) [ CONTAINS (. , "M") ]<br>RETURN $gs)<br></result></td></tr>
<tr><td>Local2 (Report)</td><td>import split as USER split;<br>FUNCTION str1($str)<br>{ split(" ", "1", $str)}<br><result><br>(FOR $gs IN document("Report.xml") //name[str1(.)]<br>WHERE str1($gs) [ CONTAINS (. , "M") ]<br>RETURN $gs)<br></result></td></tr>
<tr><td>Local3 (Thesis)</td><td>import split as USER_split;<br>FUNCTION str1to2($str)<br>{ split(" ", "1to2", $str) }<br><result><br>(FOR $gs IN document("Thesis.xml") //author[/first]<br>WHERE $gs[ CONTAINS (. , "M") ]<br>RETURN $gs) |<br>(FOR $gs IN document("Thesis.xml") //author[NOT(/first) AND NOT(/last)][str1to2(.)]<br>WHERE str1to2($gs) [ CONTAINS (. , "M") ]<br>RETURN $gs)<br></result></td></tr>
</table>

**Fig. 5.** A global query and the generated local queries

The type attribute in local is for mapping kind; 0, 1, and 2 for one-to-one, one-to-many, and many-to-one respectively; if operation attributes are included, the value of 'operation' attribute is applied to the content of the relevant local nodes in order to get data consistent with the global schema.

The <local> and <global> elements are absolute paths from the root node, which represented as '/', to the leaf nodes. For the example in Fig. 2, the <global> element for 'street' node is '/guide/restaurant/address/street' and its <local> node is '/agency2/restaurant/location'. Therefore, the mapping node for 'street' is 'location' only if the parent node of 'street' node is mapped, otherwise, its mapping node is the difference of the path between the nearest mapped ancestor node and the current node in DDXMI. This means that 'street' is mapped to *strdiff*('/agency2/restaurant/location' - 'agency2/restaurant') = 'location'. Since the mapping type of 'street' node is *2* and the attribute value of its operation is '*cstr1*', it can be easily seen that this is a 1:N mapping, and the transformation function '*cstr1*' is applied to the value of 'location' node, where '*cstr*1' is the name of function separating a string into a set of strings delimited by comma. When the 'street' node is encountered in the parsing process, it is automatically replaced by either the string '*cstr1*(location)' or 'location' depending on the type of the Quilt statement.

The mapping information for N:1 mapping types is stored in DDXMI by separating the node paths by comma. In Fig. 2, the <local> elements of 'zipcode' are '/agency2/restaurant/location/state-code' and '/agency2/restaurant/location/zip-code' and the attribute value of its operation is '0, 1', which is used to indicate the merging sequence. Therefore, 'zipcode' is transformed into '*mergepath*(state-code, zip-code)'.

### 4.2 Experimental Results

To demonstrate how the system works, we report an experiment with integration of information from 3 local documents: '*thesis*', '*reports*', and '*journal*' semi-structured documents. Assume that we are going to build 'article' database as a virtual global document from information maintained by 'thesis', 'reports', and 'journal' local documents.

The mapping between global and local schemas is shown in Fig. 4. An example Quilt queries getting author's name whose first name contains 'M' letter and the generated local queries from them are shown in Fig. 5.

## 5 Conclusion and Remaining Issues

A system for generating local queries corresponding to a query of the virtual global schema over distributed semistructured data has been described, with a focus on resolving both structural and semantic conflicts among data sources. It consists of mapping, path substitution, and node identification mechanisms. Especially, it handles the multiple mapping on an element and the node identification among the elements with the same label and different meanings.

The DDXMI file is generated by collecting the paths with the same index numbers. Global queries from end users are translated to appropriate queries to local documents by looking up the corresponding paths and possible semantic functions in the DDXMI, and node identification information. Finally local queries are executed by Kweelt.

There are several obvious limitations with the query processing algorithm and its implementation. Firstly, we extract path trees for documents without explicit schemas using an algorithm that may produce extremely large path trees for irregular semistructured data, which may be too difficult for human to handle. It is desirable to explore how to balance the accuracy and size of approximate typing in practice. Secondly, JOINs among local data are not considered. In order to fully use knowledge of the local documents for query decomposition and optimization, it is planned to extend the mapping description power to support describing and using more sophisticated kinds of relationship, and also relationships at more levels, such as local path vs. local path, document vs. document, and document vs. path.

## References

1. S. Abiteboul. Querying semistructured data. In Proceedings of ICDT, 1997
2. Peter Buneman. Tutorial: Semistructured data. In Proceedings of PODs, 1997.
3. Andrea Cal'y, Diego Calvanese, Giuseppe De Giacomo and Maurizio Lenzerini. View-based query answering and query containment over semistructured data. In: Proc. of DBPL 2001.

4. Lucian Popa, Mauricio. Hernandez, Yannis Velegrakis, Renee J. Miller, Felix Naumann, and Howard Ho. Mapping xml and relational schemas with clio. Demo on ICDE 2002.
5. Lucian Popa, Yannis Velegrakis, Renee Miller, Mauricio. Hernandez, and Ronald Fagin. Translating web data. Proc. 28th VLDB Conf., 2002.
6. Joseph Goguen. Data, schema and ontology integration. In Walter Carnielli, Miguel Dionisio, and Paulo Mateus, editors, Proc. Comblog'04, pages 21-31, 2004.T
7. D. Chamberlin, J. Robie and D. Florescu. Quilt: An XML Query Language for Heterogeneous Data Sources. Proceedings of WebDB 2000 Conference, in Lecture Notes in Computer Science, Springer-Verlag, 2000.
8. An-Hai Doan, Pedro Domingos and Alon Halevy. Reconciling schemas of disparate data sources: A machine-learning approach. Proc. SIGMOD, 2001.
9. An-Hai Doan. Thesis: Learning to Translate between Structured Representations of Data.University of Washington, 2003.
10. Hong-Hai Do and Erhard Rahm. Coma - a system for flexible combination of schema matching approaches. Proc. 28th VLDB Conf., 2002.
11. Bin He and Kevin Chen-Chuan Chang. Statistical Schema Matching across Web Query Interfaces. Proc. SIGMOD, 2003.
12. A. Y. Levy. Answering Queries Using Views: A Survey. VLDB Journal, 2001.
13. J. McHugh, S. Abiteboul, R. Goldman, D. Quass, and J. Widom. Lore: A database management systems for semistructured data. SIGMOD Record, 26, 1997.
14. Alon Levy. The Information Manifold approach to Data Integration. IEEE Intelligent Systems, vol.13, pages:12–16,1998.
15. Jayant Madhavan,Philip Bernstein and Erhard Rahm. Generic Schema Matching with Cupid. Proc. 27th VLDB Conference, 2001.
16. Sergey Melnik, Hector Garcia-Molina and Erhard Rahm. Similarity Flooding: A Versatile Graph Matching Algorithm and its Application to Schema Matching. Proc. ICDE,2002.
17. Sergey Melnik, Erhard Rahm and Philip Bernstein. Rondo: A Programming Platform for Generic Model Management. Proc.SIGMOD,2003.
18. J. D. Ullman. Information integration using logical views. International Conference on Database Theory (ICDT), pages 19-40, 1997.
19. S. Nestorov, S. Abiteboul, and R. Motwani. Inferring Structure in Semistructured Data. In Proceedings of the Workshop on Management of Semistructured Data, 1997.
20. Svetlozar Nestorov, Serge Abiteboul and Rajeev Motwani. Extracting schema from semistructured data. In Proceedings of SIGMOD, pages 295-306, 1998.
21. Y. K. Nam, J. Goguen, and G. Wang. A Metadata Integration Assistant Generator for Heterogeneous Distributed Databases. Springer, LNCS, Volume 2519, pages 1332-1344, 2002.
22. Svetlozar Nestorov and Jeffrey D. Ullman and Janet L. Wiener and Sudarshan S. Chawathe. Representative Objects: Concise representations of Semistructured, Hierarchical Data. Proceeding of ICDE, pages 79-90, 1997.
23. Erhard Rahm and Philip Bernstein. On Matching Schemas Automatically. Technical report, Dept. Computer Science, Univ. of Leipzig, 2001.
24. Li Xu and David Embley. Using Domain Ontologies to Discover Direct and Indirect Matches for Schema Elements. Proc. Semantic Integration Workshop, 2003.
25. Hector Garcia-Molina, Yannis Papakonstantinou, D. Quass, Anand Rajarman, Y. Sagiv, Jeffrey Ullman, Vasilis Vassalos and Jennifer Widom. The TSIMMIS Approach to Mediation: Data Models and Languages. Intelligent Information System, 8(2), 1997.

# Using Ontologies for Semantic Query Optimization of XML Database

Wei Sun and Da-Xin Liu

College of Computer Science and Technology, Harbin Engineering University, Harbin Heilongjiang Province, China
sunwei78@hrbeu.edu.cn

**Abstract.** As XML has gained prevalence in recent years, the management of XML compliant structured-document database has become a very interesting and compelling research area. Effective query optimization is crucial to obtaining good performance from an XML database given a declarative query specification because of the much enlarged optimization space. Query rewriting techniques based on semantic knowledge have been used in database management systems, namely for query optimization。 The main goal of query optimization is to rewrite a user query into another one that uses less time and/or less resources during the execution. When using those query optimization strategies the transformed queries are equivalent to the submitted ones. This paper presents a new approach of query optimization using ontology semantics for query processing within XML database. In fact, our approach shows how ontologies can effectively be exploited to rewrite a user query into another one such that the new query provides equally meaningful results that satisfy the intention of the user. Based on practical examples and their usefulness we develop a set of rewriting rules. In addition, we prove that the results of the query rewriting are semantically correct by using a logical model.

## 1 Introduction

Recently, XML has emerged as the *de-facto* standard for *publishing* and *exchanging* data on the Web. Many data sources export XML data, and publish their contents using DTD's or XML schemas. Thus, independently of whether the data is actually stored in XML native mode or in a relational store, the view presented to the users is XML-based. The use of XML as a data representation and exchange standard raises new issues for data management.

A large number of research approaches have used semantic knowledge for supporting data management to overcome problems caused by the increasing growth of data in local databases, and the variety of its format and model in distributed databases. The use of semantic knowledge in its various forms including meta-models, semantic rules, and integrity constraints can improve query processing capabilities by transforming user queries into other semantically equivalent ones, which can be answered in less time and/or with less resources. Known as semantic query optimization (SQO), has generated promising results in deductive, relational and object databases. Naturally, it is also expected to be an optimization direction for XML query processing.

R. Nayak and M.J. Zaki (Eds.): KDXD 2006, LNCS 3915, pp. 64 – 73, 2006.

Among the three major functionalities of an XML query language, namely, pattern retrieval, filtering and restructuring, only pattern retrieval is specific to the XML data model. Therefore, recent work on XML SQO techniques [1,2,3] focuses on pattern retrieval optimization. Most of them fall into one of the following two categories:

1. Query minimization: For example, Query tree minimization [1,3] would simplify a query asking for "all auctions with an initial price" to one asking for "all auctions", if it is known from the schema that each auction must have an initial price. The pruned query is typically more efficient to evaluate than the original one, regardless of the nature of the data source.

2. Query rewriting: For example, "query rewriting using state extents" [2] assumes that indices are built on element types. In persistent XML applications, it is practical to preprocess the data to build indices. However, this is not the case for the XML stream scenario since data arrives on the fly and usually no indices are provided in the data.

Currently, research work on the Semantic Web and data integration are focusing on using ontologies as semantic support for data processing. Ontologies have proven to be useful to capture the semantic content of data sources and to unify the semantic relationships between heterogenous structures. Thus, users should not care about where and how the data are organized in the sources. For this reason, systems like OBSERVER and TAMBIS allow users to formulate their queries over an ontology without directly accessing the data sources. In this paper, we present a new approach on how to improve the answers of queries based on semantic knowledge expressed in ontologies. Given an XML database, we assume the existence of an ontology which is associated with the database and which provides the context of its objects. We show how ontologies can be exploited effectively to rewrite a user query such that the new query can provide more "meaningful" results meeting the intention of the user.

## 2 Related Works

Work related to rewrite user query using semantic knowledge has emerged in two different research areas: Semantic query optimization and global information processing area.

**Semantic query optimization.** The basic idea of semantic query optimization (SQO) is to rewrite a query to another more efficient query, which is semantically equivalent, i.e. provides the same answer. Here, SQO approaches use semantic knowledge in various forms including semantic rules and range rules. Range rules states facts about the range of values of a given attribute, whereas semantic rules define the regularity of data for a given database. Therefore, these rules can be driven from the non-uniform distribution of values in a database. Expressing semantics in the form of horn clause sets allows the optimizer to make possible reformulations on an input query involving the insertion of new literals, or the deletion of literals, or the refuting the entire query. Several approaches on SQO have been developed to address different aspects of query processing: In [11] semantic rules have been used to derive useful information, which can reduce the cost of query plans. In [12, 13] algorithms have been developed for optimizing conjunctive sub-queries. To this end, learning techniques have been applied to generate semantic (operational) rules from a database automatically [14]. While the previous approaches are based on extracting semantic knowledge from the underlying database, current research approaches use knowledge from additional source [15, 16].

**Ontology.** The term "Ontology" or "Ontologies" is becoming frequently used in many contexts of database and artificial intelligence researches. However, there is not a unique definition of what an ontology is [7-10]. An initial definition was given by Tom Gruber: *"an ontology is an explicit specification of a conceptualization"* [7]. However, this definition is general and remains still unsatisfied for many researchers. In [8] Nicola Guarino argues that the notion of "conceptualization" is badly used in the definition. We note that many real-world ontologies already combine data instances and concepts [9]. Our definition differ from this point of view as we show later . Informally, we define an ontology as an intentional description of what is known about the essence of the entities in a particular domain of interest using abstractions, also called *concepts* and the *relationships* among them.

**Semantic query optimization of XML data.** The diversity of the XML queries (referred to in this paper as *structural queries*) results from the diversity of possible XML schemas (also called *structural schemas*) for a single conceptual model. In comparison, the schema languages that operate on the conceptual level (called *conceptual schemas*) are *structurally flat* so that the user can formulate a determined query (called *conceptual query*) without considering the structure of the source. There are currently many attempts to use *conceptual schemas* [4, 5] or *conceptual queries* [6] to overcome the problem of structural heterogeneities among XML sources.

**Contributions.** In brief, we make the following contributions in this paper: We propose an approach for using ontologies based graph model to represent semantic information of heterogeneous XML sources. This model integrates semantic information both the XML nesting structure and the domain content. These ontologies are processed lossless with respect to the nesting structure of the XML document. Finally, we describe an id-concept rule for rewriting XML query based on semantic information. The optimization is based on a mapping model based on ontology and the rules of rewriting the query on the XML sources.

## 3 The Problem Representation

Using semantic knowledge to optimize query has generated promising results in deductive, relational and object databases. Naturally, it is also expected to be an optimization direction for XML database query processing. Therefore, recent work focuses on XML optimization techniques based on semantic. It is becoming a crucial problem, how to represent the semantic information of XML database. The result is a set of semantically constrained axioms and semantically constrained relations between axioms. When a query is given to the system, the semantic transformation phase uses these stored semantic constrained sets to generate semantically equivalent queries that may be processed faster than the original query. In Fig.1, there is one DTD of XML data, which will be used as follow.

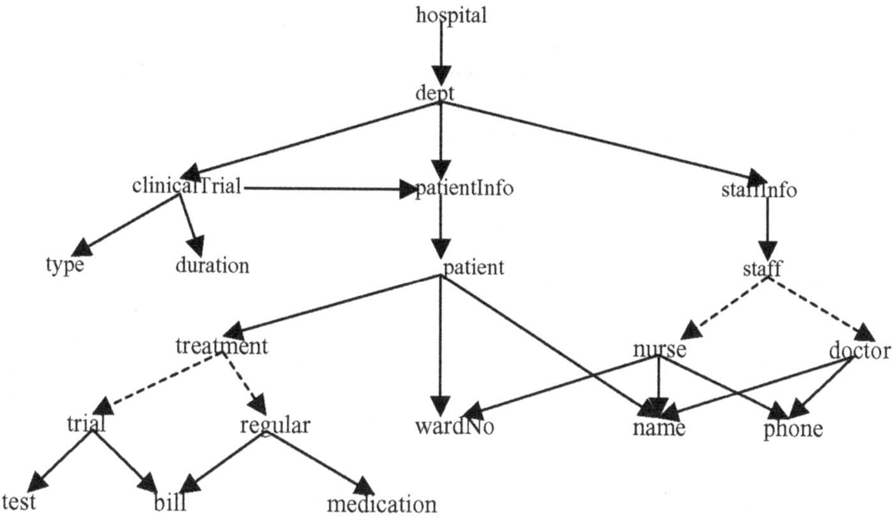


**Fig. 1.** The DTD of XML document

## 4 XML Semantic Model (XSM)

In this section, we propose the model of XML semantic, which is represented by ontologies about the content of XML document and schema. The model XSM can transform a normal query to a a semantically equivalent query, and the equivalent query has less time than the origin one to be processed.

### 4.1 Ontology Definition

Informally, we define an ontology as an intentional description of what is known about the essence of the entities in a particular domain of interest using abstractions, also called *concepts* and the *relationships* among them. Basically, the hierarchical organization of concepts through the inheritance .( . ) relationship constitutes the backbone of an ontology. Other kinds of relationship like part-whole.( . ) or synonym.( . ) or application specific relationships might exist. To the best of our knowledge, there is no work until now addressing the issue of using ontology relationships at the database instance level. Despite the disagreement upon a common meaning of an "ontology", the role of ontologies that must play is clear: Ontologies should provide a concise and unambiguous description of concepts and their relationships for a domain of interest. Ontologies are shared and reused by different agents i.e. human or/and machines.

Formally, we define an ontology as a set $\aleph$ and a set $\Re$ as follows: $\aleph = \{c_1, \ldots, c_n\}$ and $\Re$ = {"  "; "  "; "  "}, where $c_i \in \aleph$ is a concept name, and $r_i \in \Re$ is the type of the binary relation relating two concepts ($c_i$ and $r_i$ are non-null strings). Other domain-specific types may also exist. At the top of the concept hierarchy we assume the existence of a universal concept, called "Anything", which represents the most general concept of an ontology. In the literature, the word "concept" is frequently used as a synonym for the word "concept name". Hence, for the design of an ontology only one term is chosen as a name for a particular concept. Further, we assume that the terms "concept" and "concept name" have the same meaning.

### 4.2 Ontology Formal Representation

This section presents a graph-based representation of an ontology. We introduce its basic formal settings, and some related operations relevant to further discussions.

**Graph oriented model.** We represent an ontology as a directed graph *G*(*V;E*), where *V* is a finite set of vertices and *E* is a finite set of edges: Each vertex of *V* is labelled with a concept and each edge of *E* represents the inter-concept relationship between two concepts. Formally, the label of a node $n \in V$ is defined by a function *N*(*n*) = $c_i$ that maps *n* to a string $c_i$ from $\aleph$. The label of an edge $e \in E$ is given by a function *T*(*e*) = $r_i$ that maps *e* to a string $r_i$ from $\Re$.

In summary, an ontology is the set $O = \{G(V,E), \aleph, \Re, N, T\}$

**Graph operations.** In order to navigate the ontology graph, we define the following sets of concepts: *Rparent*, *DESC*, *SUBT*, *SY Ns*, *PARTs* and *WHOLEs*. We need these operations to identify nodes in the graph, which hold concepts that are of interest for our query reformulations.

Let $P_{ths}(n_1 - n_2)$ be a set of directed paths between two nodes $n_1$ and $n_2$. We denote by node(c) the node labelled by a concept c, and by child(n) and parent(n) the child-node and parent-node of a node n, respectively. Given two nodes $n_1 = node(c_1)$ and $n_2 = node(c_2)$ the operation are formulated as follows:

Rparent(r, $c_1$)=$c_2$ iff $n_2$=parent( $n_1$ ) and T[( $n_2$ , $n_1$ )]=r

DESC(r,c)={ $s \in \aleph \mid \exists p \in P_{ths}(node(c) - node(s)) : \forall e \in p, T(e) = r$ }

SYNs(c)={ $s \in \aleph \mid \exists p \in P_{ths}(node(c) - node(s)) : \forall e \in p, T(e) = "SynOf"$ }

SUBT(c)={ $s \in \aleph \mid \exists p \in P_{ths}(node(c) - node(s))$ }

Informally, *Rparent*(*r; c*) returns the label of the parent node of a concept c by following an edge of type r. *DESC*(*r; c*) returns the set of all concepts in *O* whose nodes are children of the node of c by means of descending edges of type r. Similarly, *SUBT*(*c*) returns all descendants of c for any edge-type and *SY Ns*(*c*) returns the set of all synonyms of c in *O*. In addition, we define an *Outgoings*(*n*) as a set of edge-types going out from a node n and *PARTs*(*c*) as the set of concepts whose nodes are related to the node(c) through the edges of type . . Here, two cases must be distinguished:

Case 1: If Outgoings(node(c)) ∋ . . then $PARTs(c) = A \cup B \cup C$, Where

$A = DESC( \quad . \quad ; \quad )$
$B = DESC( \quad ; \quad ), a \in A$
$C = SYNs(h) \cup SYNs(l), h \in A$ and $l \in B$

Informally, *PARTs*(*c*) is the set of concepts obtained by retrieving the labels of all nodes that are -children of the node(c) together with their - descendants and synonyms.

Case 2: If Outgoings(node(c)) ∋ . then $PARTs(c) = PARTs(s_i)$, where

$s_i \in A$ and $\forall(s_1, s_2) \in A^2$ $PARTs(s_1) = PARTs(s_2)$, $A = DESC( \quad ; \quad )$.

Informally, *PARTs* of a concept c is defined recursively in terms of its sub-concepts. It is equal to the *PARTs* of one of its sub-concepts (if they have the same *PARTs*).

Inversely, we define *WHOLEs* of a given concept c as the set of concepts $c_i$ such that $c \in PARTs(c_i)$.

### 4.3 XML Semantic Model

The XML semantic model is stated as an extension of the given ontology, denoted by $O^*$, which includes new concepts and additional relationship-types. The new concepts represent relation names, entity names, attribute names and values of the database unless they already exist. We denote these concepts by $NC_R$ $NC_E$ $NC_A$ $NC_V$, respectively. Furthermore, we call *id-concepts* the concepts that represent id-values of the database. The additional relationships have to relating these concepts to the existing ones or to each other. Their types are defined as follows:

. is the type of relationship that relates each value-concept to its associated attribute-concept or entity-concept.

. is the type of relationship between entity-concepts and attribute-concepts.

. is the type of relationship that relates an Id-concept to its associated entity-concept.

. is the type of relationship that relates entity-concepts to each other, which are associated with a particular tuple.

"RelateTo" is the type of relationship that relates relation-concepts to entity-concepts, one relation-concept with one or more entity-concepts.

"OnlyA" is the type of relationship that relates entity-concepts to each other, which are associated with an entity-concept only.

In summary, $O^*$ is defined as a set $O^* = \{G^*, \aleph^*, \Re^*, N, T\}$, where $\aleph^* = NC_E \cup NC_A \cup NC_V \cup NC_R$, and $\Re^* = \Re \cup$ {" ","  ","  ","  ", "RelateTo"}. Such as Fig.2.

**Table 1**. XPath expressions and Concepts

| XPath expressions | Concept expressions |
|---|---|
| hospital | hospital |
| hospital\dept | dept |
| Hospital\dept\clinicalTrial\patientInfo | clinical-patientInfo |
| hospital\dept\patientInfo | Non-clinical-patientInfo |
| … | … |
| hospital\dept\patientInfo\patient\treatment\regular\medication | medication |

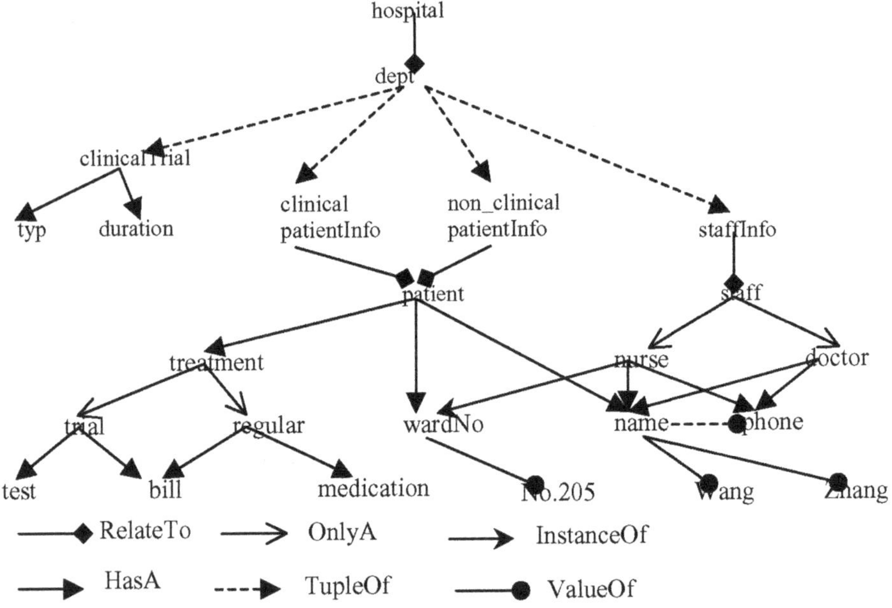


**Fig. 2.** Shows a portion of the semantic model related to DTD shown in Fig.1

**Logical Interpretation.** By using the First Order Language (FOL) the semantic model $O^*$ is defined as a theory $\Gamma$ which consists of an Interpretation $I$ and a set of well formed formulas [12]. $I$ is specified by the set of individuals $\aleph^*$ and an interpretation function $^I$. In the following, we describe the interpretation of $O^*$.

Let $n_1$ and $n_2$ be the nodes of two concepts $a$ and $b$, respectively. Formally, $\Gamma$:

$$I = (\aleph^*;\ ^I)$$

$$ISA^I = \{(a,b) \in \aleph^{*2} \mid T(n_1,n_2) = "ISA"\}$$

$SYN^I = \{(a,b) \in \aleph^{*2} \mid T(n_1, n_2) = "SynOf"\}$
$PARTOF^I = \{(a,b) \in \aleph^{*2} \mid T(n_1, n_2) = "PartOf"\}$
$HASA^I = \{(a,b) \in \aleph^{*2} \mid T(n_1, n_2) = "HasA"\}$
$VALUEOF^I = \{(a,b) \in \aleph^{*2} \mid T(n_1, n_2) = "ValueOf"\}$
$INSOF^I = \{(a,b) \in \aleph^{*2} \mid T(n_1, n_2) = "Ins\tan ceOf"\}$
$Key^I = \{a \in \aleph^{*} \mid \exists b.T(a,b) = "Ins\tan ceOf"\}$
$TUPOF^I = \{(a,b) \in \aleph^{*2} \mid T(n_1, n_2) = "TupleOf"\}$
$RELATETO^I = \{(a,b) \in \aleph^{*2} \mid T(a,b) = "\mathrm{Re}lateTo"\}$
$WHOLE^I = \{a \in \aleph^{*} \mid \forall b_1 b_2 c.ISA(a,b_1) \wedge ISA(a,b_2) \wedge PARTOF(b_1,c) \rightarrow PARTOF$
$\forall x.ISA(x,x)$
$\forall x.SYN(x,x)$
$\forall x.PARTOF(x,x)$
$\forall xyz.ISA(x,y) \wedge ISA(x,z) \rightarrow ISA(x,z)$
$\forall xy.SYN(x,y) \leftrightarrow SYN(y,x)$
$\forall xyz.SYN(x,y) \wedge SYN(x,z) \rightarrow SYN(x,z)$
$\forall xyz.ISA(x,y) \wedge SYN(y,z) \leftrightarrow ISA(x,z)$
$\forall xy\exists z.VALUEOF(x,y) \rightarrow HASA(z,y)$
$\forall xy\exists z.TUPVAL(x,y) \rightarrow INSOF(y,z)$
$\forall xyz.PARTOF(x,y) \wedge SYN(y,z) \leftrightarrow PARTOF(x,z)$
$\forall xyz.PARTOF(x,y) \wedge ISA(y,z) \leftrightarrow PARTOF(x,z)$
$\forall xyz.PARTOF(x,y) \wedge PARTOF(x,z) \leftrightarrow PARTOF(x,z)$
$\forall xyz.VALUEOF(y,z) \wedge ISA(x,y) \rightarrow VALUEOF(x,z)$
$\forall xyz.VALUEOF(y,z) \wedge SYN(x,y) \rightarrow VALUEOF(x,z)$
$\forall xyz\exists w.INSOF(x,y) \wedge HASA(y,z) \rightarrow TUPVAL(x,w) \wedge VALUEOF(w,z)$
$\forall xyz.WHOLE(x) \wedge ISA(x,y) \wedge PARTOF(y,z) \leftrightarrow PARTOF(x,z)$ $x$; $y$; $z$; $w$ are variables.

## 5 Id-Concept Rule and Validations

We note that a common feature of the rules is that after applying a rule to a query Q, the results of the reformulated query might increase. We denote by $S_Q$ and $S_{Q'}$ the result set of Q and Q', respectively. This augmentation is not arbitrary but it is proved by the semantic model $O^*$. According to $O^*$, each tuple-identifier in $S_{Q'}$ is represented by an id-concept, which is related to value-concepts through the relationship and a relation-concept through the . and relationship, respectively. $O^*$ interprets the reformulation results of a given rule as the existence of additional value-concepts, which are semantically related to those representing terms in the condition of $Q$. For brevity, we describe only an example of validation of the proposed rules using the available logical expressions from $\Gamma$.

Concerning this rule the $S_Q$ -identifiers are formally expressed by the following set of individuals $\Omega_1$, we obtain the set of individuals from Q which represents all id-concepts of the tuples in $S_{Q}$. Formally,

$$\Omega_1 = \{x \mid \exists z \forall a VALUEOF(z,a) \wedge TUPOF(z,x) \wedge INSOF(x, NC_E) \rightarrow TUPOF(z, NC_E) \wedge [ISA(NC_V, a) \vee SYN(NC_V, a)]\}.$$

## 6 Conclusions

Recently, there is a growing interest in ontologies for managing data in database and information systems. In fact, ontologies provide good supports for understanding the meaning of data. They are broadly used to optimize query processing among the distributed sources. In this paper, we use ontologies within XML database and present a new approach of query optimization using semantic knowledge from a given ontology to rewrite a user query in such way that the query answer is more meaningful to the user. To this end, we propose a set of query rewriting rules and illustrate their effectiveness by some running examples. Although these rules might not be ideal, we hope that they can bring more insight into the nature of query answers. Our approach is appropriate for database applications, where some attributes are enumerated from a list of terms. In the future, we will develop additional rewriting rules and intend to address the problem of how to establish mapping information between the database objects and ontological concepts present in an ontology associated with a specific database.

## References


1. S. Amer-Yahia, S. Cho, L. V. Lakshmanan, and D. Srivastava. Minimization of Tree Pattern Queries. *In Proc. of SIGMOD*(2001) 497–508
2. M. F. Fernandez, D. Suciu. Optimizing Regular Path Expressions Using Graph Schemas. *In Proc. of ICDE* (1998) 14–23
3. Z. Chen, H. Jagadish and L.V.S. Lakshmanan et al. From Tree Patterns to Generalized Tree Patterns; On Efficient Evaluation of XQuery. *In Proc. of 29th VLDB* (2003) 237-248
4. B. Amann, C. Beeri, I. Fundulaki, and M. Scholl. Ontology-Based Integration of XML Web Resources. *In Proceedings of the 1st International Semantic Web Conference (ISWC 2002)* 117–131

5. B. Amann, I. Fundulaki, M. Scholl, C. Beeri, and A. Vercoustre. Mapping XML Fragments to Community Web Ontologies. *In Proceedings of the 4th International Workshop on the Web and Databases (WebDB 2001)* 97–102
6. S. D. Camillo, C. A. Heuser, and R. S. Mello. Querying Heterogeneous XML Sources through a Conceptual Schema. *In Proceedings of the 22nd International Conference on Conceptual Modeling (ER2003* 186–199
7. Gruber, T.: A translation approach to portable ontology specifications. *In: Knowledge Acquisition,* 5( 2) (1993) 199-220
8. Guarino, N., Giaretta, P.: Ontologies and knowledge bases: towards a terminological clarification. *In: Knowledge Building Knowledge Sharing,ION Press.* (1995) 25-32
9. Noy, N., Hafner, C.D.: The state of the art in ontology design. *AI Magazine* 3(1997) 53-74
10. Chandrasekaran, B., Josephson, J., Benjamins, V.: What are ontologies, and why do we need them? *In: IEEE Intelligent Systems*, (1999) 20-26
11. Hsu, C., Knoblock, C.A.: Semantic query optimization for query plans of heterogeneous multidatabase systems. *Knowledge and Data Engineering,* 12 (2000) 959-978
12. Yu, C.T., Sun, W.: Automatic knowledge acquisition and maintenance for semantic query optimization. IEEE Trans. *Knowledge and Data Engineering,* 1 (1989) 362-375
13. Sun, W., Yu, C.: Semantic query optimization for tree and chain queries. *IEEE Trans. on Data and Knowledge Engineering* 1 (1994) 136-151
14. Hsu, C.: Learning effective and robust knowledge for semantic query optimization (1996)
15. Peim, M., Franconi, E., Paton, N., Goble, C.: Query processing with description logic ontologies over object-wrapped databases. technical report, University of Manchester (2001)
16. Bergamaschi, S., Sartori, C., Beneventano, D., Vincini, M.: ODB-tools: A description logics based tool for schema validation and semantic query optimization in object oriented databases. *Advances in Artificial Intelligence, 5th Congress of the Italian Association for Artificial Intelligence,* Rome, Italy (1997) 435-438

# The Expressive Language ALCNHR+K(D) for Knowledge Reasoning*

Nizamuddin Channa[1,2] and Shanping Li[1]

[1] College of Computer Science, Zhejiang University,Hangzhou, P.R. China 310027
[2] Institute of Business Administration, University of Sindh, Jamshoro, Pakistan 71000
nchanna68@yahoo.com, shan@cs.zju.edu.cn

**Abstract.** The Expressive Language ALCNHR+(D) provides conjunction, full negation, quantifiers, number restrictions, role hierarchies, transitively closed roles and concrete domains. In addition to the operators known from ALCNHR+, a restricted existential predicate restriction operator for concrete domains is supported. In order to capture the semantic of complicated knowledge reasoning model, the expressive language ALCNHR+K(D) is introduced. It cannot only be able to represent knowledge about concrete domain and constraints, but also rules in some sense of closed world semantic model hypothesis. The paper investigates an extension to description logic based knowledge reasoning by means o f decomposing and rewriting complicated hybrid concepts into partitions. We present an approach that automatically decomposes the whole knowledge base into description logic compatible and constraints solver. Our arguments are two-fold. First, complex description logics with powerful representation ability lack effectively reasoning ability and second, how to reason with the combination of inferences from distributed heterogeneous reasoner.

## 1 Introduction

Description logics (DLs) [1] are a family of logical formalisms that originated in the field of artificial intelligence as a tool for the representation of conceptual knowledge. Since then, DLs have been successfully used in a wide range of application areas such as knowledge representation, reasoning about class based formalisms (e.g conceptual database models and UML diagrams) and ontology engineering in the context of semantic web [2]. The basic syntactic entities of description logics are concepts, which are constructed from concept names (unary predicates) and role names (binary relations). Furthermore, a set of axioms (also called Tbox) are used for modeling the terminology of an application Knowledge about specific individuals and their interrelationships is modeled with a set of additional axioms (so-called ABox). Using different constructors defined with a uniform syntax and unambiguous semantics, complex concept definitions and axioms can be built from simple components. Therefore, DLs are particularly appealing both to represent ontological knowledge and to

* The research is funded by Natural Science foundation of China (No. 60174053, No. 60473052).

R. Nayak and M.J. Zaki (Eds.): KDXD 2006, LNCS 3915, pp. 74–84, 2006.

reason with it. Unfortunately, Due to the inherent complexity with the product knowledge, the expressive power needed to model complex real-world product ontologies is quite high. Practical product ontology not only needs to represent abstract concept in the application, but also the concrete domain and constrains roles [3]. Even in some scene, such as expert system, procedural rules also need to be considered. During the last few years, much research has been devoted to the development of more powerful representation system in DL family [4] [5] [6]. Despite the diversity of their representations, most of them are based on ALC [7] and its expressive successors SHIQ [8], extend the original tableau-based algorithm in different ways. It has been proved that reasoning about extensions of ALC with concrete domains is generally intractable. This problem can be moderated only if suitable restrictions are introduced in the way of combining concept constructors [9]. Homogeneous reasoning systems (or systems with homogeneous inference algorithms) have encountered the difficulty of finding the right 'trade-off' between expressiveness and computational complexity. To take advantage of the DLs popularity and flexibility in the context of semantic web, we argue that consistent DLs representation pattern is necessary. But for reasoning ability, we need to decompose the product ontology into partitions, so that different reasoning paradigms can be jointly used. The benefits of such an approach in the context of ontology sharing through the articulation of ontology interdependencies are highlighted in [10].

The rest of this paper is organized as follows: Section 2 presents the overview of the expressive language ALCNHR+(D) and section 3 Concept definitions of *ALCNHR+K(D)* knowledge base. Section 4 describes System architecture for knowledge reasoning in detail. Section 4 draws the conclusion and future work.

## 2 Overview of the Expressive Language ALCNHR+(D)

In this section, we introduce the expressive language ALCNHR+(D)[11], which support practical modeling requirements and had been implemented in the RACER (Reasoner for ABoxes and Concept Expression Reasoner) system [12]. Based on ALCNHR+(D), we further extend it by epistemic operator to capture rule knowledge in product data. The following is it's main syntax and semantics explanation. We briefly introduce the syntax and semantics of the expressive language *ALCNHR+(D)*. We assume five disjoint sets: a set of concept names $C$, a set of role names $R$, a set of feature names $F$, a set of individual names $O$ and a set of concrete objects $O_C$. The mutually disjoint subsets $P$ and $T$ of $R$ denote non-transitive and transitive roles, respectively $(R = P \cup T)$. For presenting the syntax and semantics of the language $ALCNHR+(D)$, a few definitions are required.

**Defination 1**(Concrete Domain). A concrete domain $D$ is a pair $(\Delta_D, \Phi_D)$, where $\Delta_D$ is a set called the domain and $\Phi_D$ is a set of predicate names. The interpretation name function maps each predicate name $P_D$ from $\Phi_D$ with arity n to a subset $P^I$ of $\Delta_D^n$. Concert objects from $O_C$ are mapped to an element of $\Delta_D$. A concrete domain $D$ is called admissible iff the set of predicate names $\Phi_D$ is closed under negation and $\Phi_D$ contains a name $T_D$ for $\Delta_D$ and the satisfiability problem $P_1^{n1}(x_{11},....x_{1n1}) \wedge ... \wedge P_m^{nm}(x_{m1},...x_{mnm})$ is decidable ( $m$ is finite, $P_i^{ni} \in \Phi_D$ , $ni$ is the arity of $P$ and $x_{jk}$ is a name for concrete object from $\Delta_D$ ) . We assume that $\perp_D$ is the negation of the predicate $T_D$.Using the definitions from above, the syntax of concept terms in $ALCNHR+(D)$ is defined as follows.

**Definition 2** (Concept Terms). Let C be a set of concept names with is disjoint form R and F. Any elements of C is a concept term. If $C$ and $D$ are concept terms, $R \in$ R is an arbitrary role, $S \in$ S is a simple role, $n, m \in$ , $n \geq 1$ and $m \geq 0$ , $P \in \Phi_D$ is a predicate of the concrete domain, $f, f_1,..., f_k \in$ F are features, then the following expressions are also concept terms:

$C \cap D$ (conjunction), $C \cup D$ (disjunction), $\neg C$ (negation), $\forall R.C$ (concept value restriction), $\exists R.C$ (concept exists restriction), $\exists_{\leq m} S$ (at most number restriction), $\exists_{\geq n} S$ (at least number restriction), $\exists f, f_1,..., f_k.P$ (predicate exists restriction), $\forall f.\perp_D$ (no concrete domain filler restriction).

**Definition 3** (Terminological Axiom, TBox). If $C$ and $D$ are concept terms, then $C \subseteq D$ is a terminological axiom. A terminological axiom is also called generalized concept inclusion or GCI. A finite set of terminological axioms is called a terminology or *TBox*. The next definition gives a model-theoretic semantics to the language introduced above. Let $D = (\Delta_D, \Phi_D)$ be a concrete domain.

**Definition 4** (Semantics). An interpretation $I_D = (\Delta_I, \Delta_D, {}^I)$ consists of a set $\Delta_I$ (the abstract domain), a set $\Delta_D$ (the domain of the 'concrete domain' $D$ ) and an interpretation function ${}^I$ . The interpretation function ${}^I$ maps each concept name $C$ to a subset $C^I$ of $\Delta_I$, each role name $R$ from R to a subset $R^I$ of $\Delta_I \times \Delta_I$. Each feature $f$ from F is mapped to a partial function $f^I$ from $\Delta_I$ to $\Delta_D$ where $f^I(a) = x$ will be written as $(a, x) \in f^I$ . Each predicate name $P$ from $\Phi_D$ with arity n is mapped to a subset $P^I$ of $\Delta_D^n$ . Let the symbols $C$ , $D$ be concept expressions, $R$ , $S$ be role names, $f, f_1,...f_n$ be features and let $P$ be a predicate name. Then, the interpretation function is extended to arbitrary concept and role terms as follows ( denotes the cardinality of a set):

$$(C \cap D)^I := C^I \cap D^I, (C \cup D)^I := C^I \cup D^I, (\neg C)^I := \Delta_I \setminus C^I$$

$$(\exists R.C)^I := \{a \in \Delta_I \mid \exists b : (a,b) \in R^I, b \in C^I\}$$

$$(\forall R.C)^I := \{a \in \Delta_I \mid \forall b : (a,b) \in R^I, b \in C^I\}$$

$$(\exists_{\geq n} R)^I := \{a \in \Delta_I \mid \| \{b \mid (a,b) \in R^I\} \| \geq n\}$$

$$(\exists_{\leq m} R)^I := \{a \in \Delta_I \mid \| \{b \mid (a,b) \in R^I\} \| \leq m\}$$

$$(\exists f_1,...,f_n.P)^I := \{a \in \Delta_I \mid \exists x_1,...,x_n \in \Delta_D$$

$$:(a,x_1) \in f_1^I,...,(a,x_n) \in f_n^I, (x_1,...,x_n) \in P^I\}$$

$$(\forall f.\bot_D)^I := \{a \in \Delta_I \mid \neg\exists x_1 \in \Delta_D : (a,x_1) \in f^I\}$$

An interpretation $I_D$ is a model of a concept $C$ iff $C^{I_D} \neq \varnothing$. An interpretation $I_D$ satisfies a terminological axiom $C \subseteq D$ iff $C^I \subset D^I$ . $I_D$ is a model of a TBox iff it satisfies all terminological axioms $C \subseteq D$ in TBox. An interpretation $I_D$ is a model of an RBox iff $R^I \subseteq S^I$ for all role inclusions $R \subseteq S$ in $R$ and, in addition, $\forall transtive(R) \in R : R^I = (R^I)^+$

**Definition 5** (Assertional Axioms, ABox). Let $O = O_O \cup O_N$ be a set of individual names (or individuals), where the set $O_O$ of "old" individuals is disjoint with the set $O_N$ "new"individuals. Old individuals are those names that explicitly appear in an ABox given as input to an algorithm for solving an inference problem, i.e. the initially mentioned individuals must not be in $O_N$. Elements of $O_N$ will be generated internally. Furthermore, let $O_C$ be a set of names for concrete objects $(O_C \cap O = \varnothing)$. If C is a concept term, $R \in R$ , a role name, $f \in F$ a feature, $a,b \in O_O$ , are individual names and $x, x_1,...x_n \in O_C$, are names for concrete objects, then the following expressions are assertional axioms or

ABox assertions:
$a : C$ (concept assertion), $(a,b) : R$ (role assertion), $(a,x) : f$ (concrete domain feature assertion) and $(x_1..x_n) : P$ (concrete domain predicate assertion).
For example, part of the product model, illustrated in figure 1, can be represented as following:

$PC \subseteq \forall has\_part.HD \cap \forall has\_part.FD \cap$

$\forall has\_part.Mother\_board \cap \forall has\_part.OS \cap \exists has\_part.HD.storag\_space,$

$has\_part.OS.storag\_space\_req.more$

$HD \subseteq \forall storage\_space.integerOS \subseteq \forall storage\_space\_requirment.integer$ .

## 2.1 Epistemic Operation K

In some system, such as computer-aided process planning (CAPP) rules are used to express knowledge, especial heuristic rules and default rules [13]. The simplest variant of such rules are expressions of the form $C \Rightarrow D$, where $C$, $D$ are concepts. Operationally, a forward process can describe the semantics of a finite set of rules. Starting with an initial knowledge base $K$, a series of knowledge bases $K^{(0)}, K^{(1)}, K^{(2)}$,.......... is constructed, where $K^{(0)} = K$ and $K^{(i+1)}$ is obtained from $K^{(i)}$ by adding a new assertion $D(a)$ whenever there exists a rule $C \Rightarrow D$ such that $K^{(i)} \models C(a)$ holds, but $K^{(i)}$ does not contain $D(a)$. These processes eventually halt if the initial knowledge base contains only finitely many individuals and there are only finitely many rules. The difference between the rule $C \Rightarrow D$ and the inclusion axiom $C \subseteq D$ is that the rule is not equivalent to its contra positive $\neg D \Rightarrow \neg C$. In addition, when applying rules one does not make a case analysis. For example, the inclusions $C \subseteq D$ and $\neg C \subseteq D$ imply that every object belongs to D, whereas none of the rules $C \Rightarrow D$ and $\neg C \Rightarrow D$ applies to an individual $a$ for which neither $C(a)$ nor $\neg C(a)$ can be proven. In order to capture the meaning of rules in a declarative way, we must augment description logics by an operator K [14], which does not refer to objects in the domain, but to what the knowledge base knows about the domain. Therefore, K is an epistemic operator.

To introduce the K-operator, we enrich both the syntax and the semantics of description logic languages. Originally, the K-operator has been defined for ALC [15]. First, we add one case to the syntax rule that allows us to construct epistemic concepts: $C, D \rightarrow \mathrm{K}C$ (epistemic concept). Intuitively, the concept $\mathrm{K}C$ denotes those objects for which the knowledge base knows that they are instances of $C$. Next, using K, we translate rules $C \Rightarrow D$ into inclusion axioms $\mathrm{K}C \subseteq D$.

For example, rules like this: "in a computer, if the motherboard type is B1, then the CPU is only limited to 386 types and the operation system is only limited to Linux can be represented as:

$K(\forall has\_part.B1) \Rightarrow \forall has\_part.linux$. And it can be translated into:

$K(\forall has\_part.B1) \subseteq \forall has\_part.linux$.

Intuitively, the K operator in front of the concept C has the effect that the axiom is only applicable to individuals that appear in the ABox and for which ABox and TBox imply that they are instances of $C$. Such a restricted applicability prevents the inclusion axiom from influencing satisfiability or subsumption relationships between concepts. In the sequel, we will define a formal semantics for the operator K that has exactly this effect. A rule knowledge base is a triple $K = (T, A, R)$, where $T$ is a TBox, $A$ is an ABox, and $R$ is a set of rules written as inclusion axioms of the form as $\mathrm{K}C \subseteq D$. The procedural extension of such a triple is the knowledge-base $\bar{K} = (T, \bar{A})$ that is obtained from $(T, A)$ by applying the trigger rules as described above. We call the extended knowledge base ALCNHR$^+$K(D) knowledge base, because it extended by the operator K. The semantics of epistemic inclusions will be defined in such a way that it applies only to individuals in the knowledge base that provably are instances of $C$, but not to arbitrary domain elements, which would be the case if we dropped K. The semantics will go beyond first-order logic because we not only have to interpret concepts, roles and individuals, but also have to model the knowledge of a knowledge base. The fact that a knowledge base has knowledge about the domain can be understood in such a way that it considers only a subset $W$ of the set of all interpretations as possible states of the world. Those individuals that are interpreted, as elements of $C$ under all interpretations in W are then "known" to be in $C$. To make this formal, we modify the definition of ordinary (first-order) interpretations by assuming that: There is a fixed countable infinite set $\Delta$ that is the domain of every interpretation (Common Domain Assumption); There is a mapping from the individuals to the domain elements that fixes the way individuals are interpreted

(Rigid Term Assumption). The Common Domain Assumption guarantees that all interpretations speak about the same domain. The Rigid Term Assumption allows us to identify each individual symbols with exactly one domain element. These assumptions do not essentially reduce the number of possible interpretations. As a consequence, properties like satisfiability and subsumption of concepts are the same independently of whether we define them with respect to arbitrary interpretations or those that satisfy the above assumptions. Now, we define an *epistemic interpretation* as a pair $(I,W)$, where $I$ is a first-order interpretation and $W$ is a set of first-order interpretations, all satisfying the above assumptions. Every epistemic interpretation gives rise to a unique mapping $^{I,W}$ associating concepts and roles with subsets of $\Delta$ and $\Delta\times\Delta$, respectively. For T, $\perp$ for atomic concepts, negated atomic concepts, and for atomic roles, $^{I,W}$ agrees with $^{I}$. For intersections, value restrictions, and existential quantifications, the definition is similar to the one of $^{I}$.

$$(C\cap D)^{I,W}=C^{I,W}\cap D^{I,W}\qquad (\forall R.C)^{I,W}=\{a\in\Delta\mid\forall b.(a,b)\in R^{I,W}\rightarrow b\in C^{I,W}\}$$

$$(\exists R.\mathrm{T})^{I,W}=\{a\in\Delta\mid\exists b.(a,b)\in R^{I,W}\}$$

For other constructors, $^{I,W}$ can be defined analogously. It would also be possible to allow the operator **K** to occur in front of roles and to define the semantics of role expressions of the form **K**$\mathcal{R}$ analogously. However, since epistemic roles are not needed to explain the semantics of rules, we restrict ourselves to epistemic concepts.

## 3 Concept Definitions of *ALCNHR+K(D)* Knowledge Base

After rules in ontology are eliminated through operator $K$, the $ALCNHR^{+}K(D)$ knowledge base only includes concept definitions, which can be decomposed into three concepts:

**Atomic concepts**, which define the ground, constructs for ontology modeling. Objects responding to atomic concepts in information system are directly implemented by basic data structure, which connect the data level and semantic level in the hierarchy of knowledge representation. For example, in figure 1, i.e. part of a computer configuration model, the concept "HD1" own an attribute "storage_space", which is inherited form the further concept, whose value is an integer value. So "storage_space" is a concrete concept.

**Abstract concepts**, which are defined through relationships/attributes declarations with hybrid concepts and other abstract concepts, such as "HD".

**Hybrid concepts,** which are defined through relationships/attributes declarations with atomic concepts and other abstract concepts or hybrid concepts, such as "HD1". To avoid the undecidable inferential problems brought by concrete domain, hybrid concepts are decomposed into an abstract one, an image concrete concept which only contains the concrete concepts and their constrains projected from the source hybrid concept. The link relationship between image concrete concept and abstract concept is implied by the name of image concrete concept. So $ALCNHR^{+}K(D)$) knowledge base denoted as $\Pi_{KB}$ can be divided into partitions as $\Pi_{DL}$, i.e. a set of DL-oriented statements which do not exceed the expressive power of the selected DL-based system, and $\Pi_{CS}$ i.e. a set of non-DL statements which contains the concrete knowledge filtered out to from $\Pi_{DL}$. As a result, instead of reasoning with constrains directly, DL-based systems provide inferential services without being aware of the existence of constraint reasoning. All the information related to concrete domains is removed form concept definitions. Thus, only the proper DL-based constructors, which are admitted by the selected DL-based inferential engines are left.

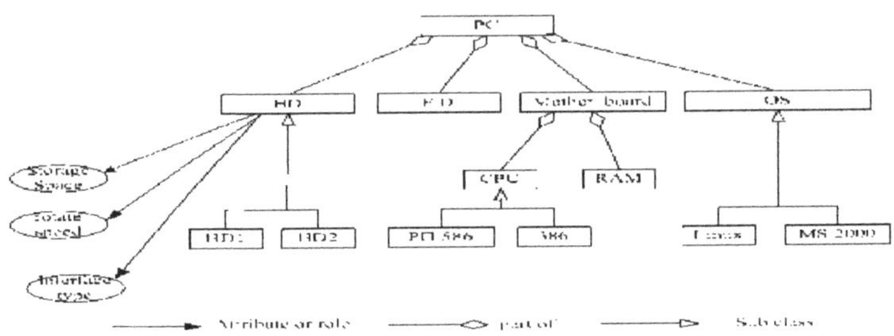


**Fig. 1.** Knowledge model for PC

For instance, let us supposed that the storage space of "HD1" type hard disk are to be required to be more than 4 GB, and the "MS 2000" need at least 2 GB storage space. In order to decompose the hybrid concept, we have

$$HD1 \subseteq Hard\_disk \cap \forall storage\_space.storage\_space_{HD1}$$

$$MS\_2000 \subseteq Operation\_System \cap storage\_space\_req.storage\_space\_req_{oper\_system}$$

In the above expression, the "storage space" restriction is replaced by an atomic concept "storage_space" which has the same name with the attribute name, but with a subscript which denote where the atomic concept comes from. Meanwhile, the restrictions on the hybrid concept is given as

$$storage\_space_{HD1} \geq 4\times 2^{30} \quad storage\_space\_req_{oper\_system} \geq 2\times 2^{30}$$

Now, by normalizing the knowledge base we split the concepts definitions and restriction into two parts. First, we replace all the hybrid concepts with the wrapper concepts, which are rewrite only by relationship or attribute with abstract concepts, and add new atomic concepts, such as "storage_apaceHD1" into the DL parts. Second, all the image concrete concepts acting as constraints variables are stored in the non-DL part together with their default domain, such as

$$\begin{array}{ll} storage\_space_{HD1} & storage\_space\_req_{oper\_system} \\ type: integer \;\ldots\ldots\ldots\;\ldots & type: integer \\ domain: \geq 4\times 2^{30} & domain: \geq 2\times 2^{30} \end{array}$$

In default, domain field is the range allowed by data type. The above statements are translated into the underlying modeling languages of the cooperative inferential engines. Subsequently, translated statements are loaded into DL and CPL inferential engines. According to the results from both inferential engines, a reasoning coordinator creates hierarchical structures of hybrid concepts, which are introduced into DL definitions through the atomic axioms concepts. In our example, after loading the non-part into an external constraints solver, we obtain a new partial order:

$storage\_space\_req_{operation\_system} \subseteq storage\_space_{HD1}$ Sending such information back to join the original DL part knowledge base, which can be used directly by DLs reasoner. We can conclude that, between satisfying other constraints, if a computer has a "HD1" type hard disk, operation system "linux" can be installed on it.

## 4 System Architecture for Knowledge Reasoning.

The STEP standard, ISO 10303, is the predominant international standard for the definition, management, and interchange of product data, being used in a wide variety of industries from aerospace, shipbuilding, oil and gas to power generation [16]. Central to the standard is the product data model, which are specified in EXPRESS (ISO 10303-11), a modeling language combing ideas from the entity-attribute-relationship family of modeling languages with object modeling concepts. To satisfy the large number of sophisticated and complex requirements put forwards by large-scale industry, the EXPRESS language has powerful expressing constructs to describe complicated product information, and had been used to build up a family of robust and time-tested standard application protocols, which had been, implemented in most Product data management (PDM) and Computer-Aided-X (CAX) systems.

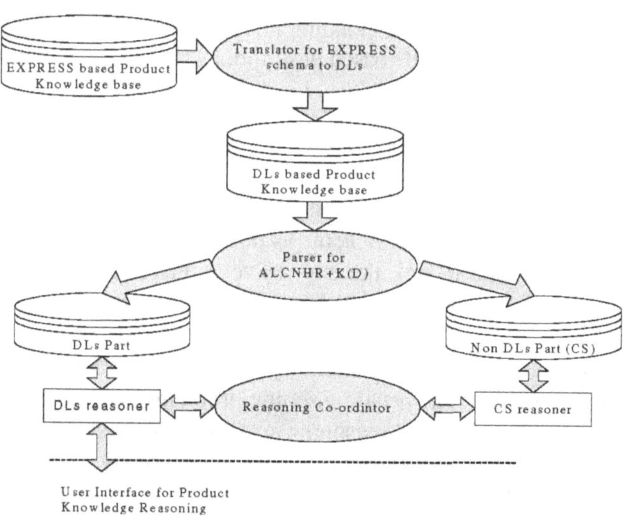


**Fig. 1.** Architecture for Knowledge Reasoning

IPDM systems manages "data about data" or metadata and provides data management and integration at the image, drawing and document levels of coarse-grain data. CAX systems have provided engineering applications with high-performance solutions.

In our former works [17] [18][19], we had proposed a translation mechanism, which rewrites the EXPRESS, based product knowledge base into DL based. So the system architecture for product data reasoning is composed of three modules, as shows in figure2.

- The translator for EXPRESS schema to DLs;
- Parser for ALCNHR+K(D), divides DLs with constraints and concrete domain to $\prod_{DL}$ and $\prod_{CS}$ sub knowledge base.
- Reasoning co-coordinator, which is the link between DLs reasoner and CS reasoner

The combined reasoning process is as follows:

1. Parse the input EXPRESS schema and translate it into the expressive DL language-ALCNHR+K(D).
2. Parse the DL based product knowledge baseextract the concrete image concepts form hybrid concepts and decompose it into homogeneous parts: DL, non-DL (the concrete value and constraints).
3. Check the consistency of constraints and propagate them in order to maintain a full path-consistency by reducing the set of possible values associated with each constrained variable.
4. Update DL-based representation with the quasi-ordering between the atomic concepts which are the corresponding image concept for each variable.
5. Update and classify the DL-based descriptions based on the new knowledge.

## 5 Conclusions and Future Work

In previous sections we presented architecture for reasoning on product knowledge, which takes originally EXPRESS Schema as input. In order to capture the semantic of complicated product data model, the expressive language ALCNHR+K (D) is introduced. It cannot only represent knowledge about concrete domain and constraints, but also rules in some sense of closed world semantic model hypothesis. To avoid the undecidable inferential problems brought by the extension. A partition based reasoning approach is proposed. The usual reasoning problems, such as concept subsuming, can be resolved by the combined reasoning systems, which take the DL reason engine as the core part. Utilizing current Semantic Web technology, product knowledge can be embedded inside Web resources. One feature of this capability is the data sources, which are readily available for consumption by a wide variety of Semantic Web users. Our proposed product knowledge reasoning architecture can be used to Semantic Web based search engines and discovery services. For further work, we need to optimize the hybrid reasoning system to adapt diverse application domain.

## References


1. Calvanese, D., De Giacomo, G., Lenzerini, M., Nardi, D., and Rosati, R. 1998. Description logic framework for information integration. In Proceedings of the 6th International Conference on rinciples of Knowledge Representation and Reasoning (KR'98). 2–13.
2. The Semantic Web lifts off 'by Tim Berners-Lee and Eric Miller, W3C. ERCIM News No. 51, October 2002
3. Felix Metzger, "The challenge of capturing the semantics of STEP data models precisely", Workshop on Product Knowledge Sharing for Integrated Enterprises (ProKSI'96), 1996.
4. F. Baader and U. Sattler, "Description Logics with Concrete Domains and Aggregation", In H. Prade, editor, Proceedings of the 13th European Conference on Artificial Intelligence (ECAI-98), pages 336-340. John Wiley & Sons Ltd, 1998.
5. F. Baader and R. Küsters, "Unification in a Description Logic with Transitive Closure of Roles". In R. Nieuwenhuis and A. Voronkov, editors, Proceedings of the 8th International Conference on Logic for Programming, Artificial Intelligence, and Reasoning (LPAR 2001), volume 2250 of Lecture Notes in Computer Science, pages 217–232, Havana, Cuba, 2001. Springer-Verlag.
6. V. Haarslev, C. Lutz, and R. Möller, "A Description Logic with Concrete Domains and Role-forming Predicates". Journal of Logic and Computation, 9(3):351–384, 1999.
7. The Description Logic Handbook, edited by F. Baader, D. Calvanese, DL McGuinness, D. Nardi, PF Patel-Schneider, Cambridge University Press, 2002.
8. Ian Horrocks, Ulrike Sattler, "Optimised Reasoning for SHIQ", ECAI 2002: 277-281.
9. I. Horrocks, U. Sattler, and S. Tobies, "Practical Reasoning for Very Expressive Description Logics". Logic Journal of the IGPL, 8(3):239–264, May 2000.
10. E. Compatangelo, H. Meisel, "K-Share: an architecture for sharing heterogeneous conceptualizations". In Intl. Workshop on Intelligent Knowledge Management Techniques (I-KOMAT'2002) - Proc. of the 6th Intl. Conf. on Knowledge-Based Intelligent Information & Engineering Systems (KES'2002), pages 1439–1443.
11. Volker Haarslev, Ralf Möller, Michael Wessel, "The Description Logic ALCNHR+ Extended with Concrete Domains: A Practically Motivated Approach". IJCAR 2001: 29-44.
12. Domazet D., "The automatic tool selection with the production rules matrix method". Annals of the CIRP, 1990, 39(1): 497~500.
13. Volker Haarselev and Ralf Moller. RACER system Description. In proceedings of the International Joint Conference on Automated Reasoning(IJCAR 2001), Volume 2083, 2001.
14. Dretske, Fred, "Epistemic Operators, The Journal of Philosophy", Vol. LXVII, No.24, Dec. 24, pp.1007-1023.
15. Donini, F. M., Lenzerini, M., Nardi, D., Nutt, W., and Schaerf, A., "Adding epistemic operators to concept languages". In Proceedings of the 3rd International Conference on the Principles of Knowledge Representation and Reasoning (KR'92). Morgan Kaufmann, Los Altos, 342–353.
16. Mike Pratt, "Introduction to ISO 10303 - The STEP Standard for Product Data Exchange", ASME Journal of Computing and Information Science in Engineering, November, 2000
17. Xiangjun Fu, Shanping Li, "Ontology Knowledge Representation for Product Data Model", Journal of Computer-Aided Design & Computer Graphics, to appear (in Chinese).

18. Xiangjun Fu, Shanping Li, Ming Guo, Nizamuddin Channa "Methodology for Semantic Representing of Product Data in XML", In Proceedings of Advance Workshop on Content Computing, LNCS, 2004
19. Nizamuddin Channa, Shanping Li, Xiangjun Fu "Product Knowledge Reasoning: A DL-based approach" In proceedings Seven International Conference on Electronic Commerce (ICEC'05) Xi'an, China PP:692-697 © ACM 2005

# A New Scheme to Fingerprint XML Data*

Fei Guo[1], Jianmin Wang[1], Zhihao Zhang[1], and Deyi Li[1, 2]

[1] School of Software, Tsinghua University, Beijing 100084, China
f-guo03@mails.tsinghua.edu.cn
jimwang@tsinghua.edu.cn
zhangzh02@mails.tsinghua.edu.cn
[2] China Constitute of Electronic System Engineering,
Beijing 100039, China
ziqinli@public2.bta.net.cn

**Abstract.** Watermarking has been used for digital rights protection over different types of contents on the Web. Since XML data is widely used and distributed over the Internet, watermarking XML data is of great interest. In this paper, we present a new watermarking scheme to embed different fingerprints in XML data. The fingerprint can be used to trace illegal distributors. We also take into consideration that XML data usually contains categorical elements which can't tolerant much modification, our scheme attempts to reduce modifications without bringing down the robustness of the fingerprint. Modifications could be reduced by choosing different patterns to insert. The experiments show that our scheme is effective to make less distortion to the original data and the fingerprint maintains the same robustness level at the same time.

## 1 Introduction

Today, mass of data could be copied and distributed on the Web easily. Since valuable data could be resold for illegal profit, it's important to claim original ownership of a redistributed copy and trace traitors as well. Watermarking is a class of information hiding techniques to protect digital contents from unauthorized duplication and distribution by introducing small errors into the object being marked. These small errors constitute a watermark or fingerprint. Fingerprinting is often discussed in comparison or extension to watermarking [5]. A watermark is used to identify the owner while a fingerprint is used to identify illegal distributor. For example, the owner embedded a unique fingerprint to each user (user1, user2, user3), see figure 1. When an unauthorized copy on the web is found, the owner could detect the user3's fingerprint to argue ownership and track back to identify user3 to be the illegal distributor out of other users.

Since XML is designed especially for applications on the Web and more and more Internet contents are defined in XML, it's of great significance to fingerprint valuable XML documents.

* This research is supported by National Natural Science Foundation of China under Project No. 60473077.

R. Nayak and M.J. Zaki (Eds.): KDXD 2006, LNCS 3915, pp. 85–94, 2006.

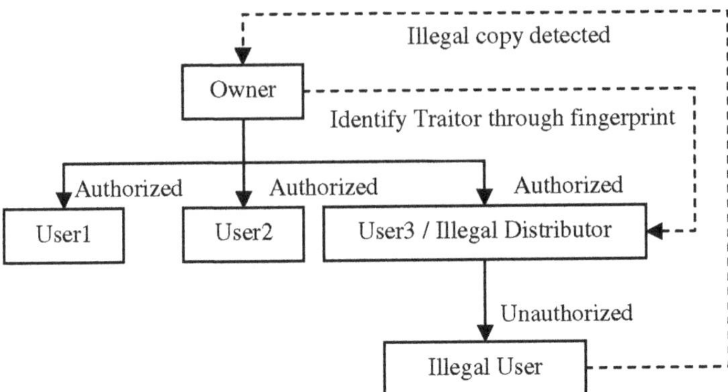


**Fig. 1.** Using a fingerprint to identify illegal distributors

Wilfred Ng and Ho-Lam Lau [4] present a selective approach of watermarking XML. It's successful to insert a watermark to prove ownership over XML data, but it can't insert different fingerprints to help identify illegal distributors. Sion [2] presents an even grouping method for fingerprinting relational data. We extend his techniques into a varied-size grouping method to fingerprint XML data. Agrawal [1] presents an effective watermarking technique for relational data to modify some bit positions of some attributes of some tuples and gives good mathematic analysis. It gives the foundation of our analysis on confidence level within each group.

In this paper, we introduce a scheme to embed fingerprints of ordered bits. We first classify elements into groups and embed one bit of the fingerprint in each group. To maintain the same grouping result for successful detection, we introduce a varied-size grouping method. A value of "remainder" of each element is calculated to identify which group it belongs to, and the ascending order of the "remainder" naturally preserves the order of the groups, also the order of the fingerprint. Thus, only the number of groups that equals the length of the fingerprint is needed to calculate the same "remainder" when detecting the fingerprint. The even grouping method [2] has to record extra classifying information for each group, which is of the same size of the fingerprint or even more and is not necessary.

All robust watermarking schemes [1] [2] [4] have to make some distortions to the original data. So it's assumed that small errors will not decrease the usability of the original data remarkably in all robust watermarking algorithms. For example, to embed a mark, number <byteCount>5440</byteCount> can be modified to <byteCount>5439</byteCount>, word <TEAM_CITY>Los Angeles</TEAM_CITY> can be replaced by its synonym <TEAM_CITY>L.A.</TEAM_CITY>. But it's hard to define what change is within the acceptable level. In many real life cases, changes tend to be too big to meet the assumption, especially for categorical data. In [7], the distortion to each selected categorical item may be significant, for example, even one bit error in such as social security number is not acceptable. However, [7] assumed that it's acceptable if only a small part of data is modified. So we attempt to find a way to minimum the part to be modified to minimum the change to the original data, meanwhile preserve the same robustness level of the fingerprint. We believe that data with fewer errors is more valuable than data with more errors, both for categorical data and numeric data. In our scheme, we use the fingerprint bit to be embedded to choose the inserting positions, thus different fingerprint bit is represented by different positions, not by the value of the selected mark positions, so we have a choice to either set the selected bit value or word value to "1" or "0", corresponding to either $pattern_1$ or $pattern_0$. Since some of the selected mark positions may meet the pattern naturally, i.e., no need to be modified, we can choose a pattern that needs the minimum modifications. In other words, we examine the original values of selected mark positions, and choose a pattern that original values tend to be, thus minimum modifications. So we don't reduce the selected bit positions to minimum modifications, which mean that we don't bring down the mark ratio, thus the fingerprint maintains the same robustness level. Our experiment shows that in some cases, we can reduce the modifications by 1/4 compared with [4] at the same mark ratio. For numeric data, it means less effect on mean and variance; for categorical data, it means we reduce the probability of destroying an element (e.g. any distortion on the social security number) to 3/4.

The rest of the paper is organized as follows: Part 2 provides our insertion algorithm and detection algorithm. Part 3 gives the implementation of our fingerprinting scheme and the analysis on modification amount and fingerprints' robustness. We conclude with a summary and directions for future work in Part 4.

## 2 Scheme to Fingerprint XML data

In this section, we provide our insertion and detection algorithms. The notations used in this paper are shown in Table 1:

**Table 1.** Notations

| | |
|---|---|
| $1/\gamma$ | Target fraction of elements marked / mark ratio |
| $\varepsilon$ | Number of candidate bits to be modified |
| $k$ | The secret key |
| $\alpha$ | Significance level for detecting a watermark |
| *PA* | Primary Data in each element |
| *N* | Number of elements in XML data |
| o | Concatenating |

### 2.1 Insertion Algorithm

A primary data (PA) used to identify each element should be predefined by the owner; also the candidate bit positions and candidate word positions *num_of_word_in_value* should be predefined. The primary data which is used as the primary key in relational databases should be unique and unchanged. For example, the <social_security_number> could be used as PA. If no such data exists, we can construct a virtual PA as stated in [5]. For example, we may use the combination of <SURNAME> and <GIVEN_NAME> instead. We use a one-way hash function value affected by the PA and the secret key k to decide the group position and mark position. Since only the owner knows the secret key, it's hard for an attacker to find our fingerprint.

First we transform a fingerprint in any form (e.g. fingerprint of a picture) into a bit flow and the length of the fingerprint should be recorded for detection. Then we calculate the remainder i for each element at line 4 in our insertion algorithm below. Then elements with same values of i and meet line 5 at the same time are collected into the same $i^{th}$ group. The $i^{th}$ bit of the fingerprint will be inserted into elements in the $i^{th}$ group. Thus we have fpt_length (bit number of the fingerprint) groups. The ascending order of i ranging from 0 to fpt_length-1 naturally preserves the order of the fingerprint. Since the hash result of MD5 we used is expected to be uniform distributed, each group may have varied but similar number of elements.

Let's see how a bit of fingerprint is inserted in each group. We use the fingerprint bit value to choose the inserting positions, see line 5, 13 and 16. It decides which element to mark, and which bit or word to be modified. The mark ratio  is used to choose the insertion granularity. Notice that the elements selected to mark and the modification position j are different when the fingerprint to be embedded is "1" from when it's "0". In subroutine pattern($subset_i$), we count the original values of each selected position within a group and choose the mark pattern. Since most categorical data is in textual form, we use the parity of the word's Hash value to represent value "1" or "0" corresponding to bit value for numeric data, see subroutine value(word). For $pattern_1$, we set each selected position value into "1", and for $pattern_0$, we set each selected position value into "0". For example, if the selected values are eight "1" and two "0" in a group, $pattern_1$ is chosen (see line 32) and only two elements have to be changed. In the opposite situation, if the selected values are eight "0" and two "1" in a group, $pattern_0$ is chosen (see line 31) and only two elements have to be changed too. Then subroutine embed($subset_i$) will modify the selected positions according to the pattern chosen, two elements in the example. How to modify the selected position for numeric and textual element is shown at line 14 and 18 respectively.

**Algorithm 1.** Insertion algorithm

```
// Only the owner of data knows the secret key k.
// R is the XML document to be marked.
// fpt_length is the length of the fingerprint embedded.
1)   fpt[] = bit(fingerprint)
2)   record fpt_length        // length of the fingerprint is recorded for detection
3)   foreach element τ ∈R do {
4)     i = Hash(PA o k) mod fpt_length        // fpt[i] to be inserted
5)     if(Hash(fpt[i] o PA o k) mod γ equals 0) then    // mark this element
6)       subset_i ← element }
7)   foreach subset_i
8)     embed(subset_i)
9)   subroutine embed(subset_i)
10)    mask[i] = pattern(subset_i)
11)    foreach element in subset_i do {
12)      if(element is numeric)
13)        j = Hash(PA o k o fpt[i]) mod ε
14)        set the j^th bit to mask[i]        // modify the j^th candidate bit
```

```
15)       else if(element is textual)
16)         j = Hash(PA o k o fpt[i]) mod num_of_word_in_value
17)         if(value( the j^th word ) is not equal to mask[i] )      // modify the j^th word
18)           replace the j^th word by a synonym s where value(s) equals mask[i]
19)         else do nothing }

20) subroutine pattern(subset_i)          // choose a pattern for less modification
21)   count_0 = count_1 = 0
22)   foreach element in subset_i do {
23)     if(element is numeric)
24)       j = Hash(PA o k o fpt[i]) mod ε
25)       if(the j^th bit equals 0) count_0 increment
26)       else count_1 increment
27)     else if(element is textual)
28)       j = Hash(PA o k o fpt[i]) mod num_of_word_in_value
29)       if(value( the j^th word )) count_0 increment
30)       else count_1 increment }
31)   if(count_0 > count_1) mask = 0     // pattern_0
32)   else mask = 1                      // pattern_1
33)   return mask

34) subroutine value(word)
35)   if(Hash( word ) is even)
36)     value = 0
37)   else value = 1
38)   return value
```

## 2.2 Detection Algorithm

To detect a fingerprint, the owner has to use the same secret key, the same predefined parameters, the fingerprint length recorded when inserting and choose a significance level for detection.

First we form similar groups, see line 3 in our detection algorithm below, thus preserve the same fingerprint order. Next we try to detect one bit of fingerprint from each group. If the embedded fingerprint is "0", compared with the insertion process, we can find exactly the same elements at line 10, and the same selected positions at line 13 and 17. For a non-marked document, because the positions are selected randomly, the probabilities of a selected position value to be either "0" or "1" are both 1/2 approximately. But for a marked document, we are expected to see that the values of each selected position are all the same, either "0" or "1", i.e., $match\_count_0$ = $total\_count_0$ or $match\_count_0 = 0$. We use the significance level set by the owner to calculate a threshold (see line 35), such that either if $match\_count_0$ is larger than threshold or is smaller than ($total\_count_0$ – threshold), we can claim that a fingerprint bit of "0" has been embedded with the confidence level of (1 - ), otherwise, we say a fingerprint bit of "0" isn't detected. Then we check if the embedded fingerprint is "1" (see line 20), the process is almost the same. If both "1" and "0" haven't be detected, we conclude that no fingerprint has been embedded at the confidence level of $(1 - \alpha)$.

**Algorithm 2.** Detection algorithm

```
// k, γ, ε and num_of_word_in_value have the same values as in watermark insertion.
// fpt_length has the same value with recorded when inserting.
// α is the significance level for detecting a fingerprint bit.
// S is a marked XML document.
foreach element τ ∈S do {
  i = Hash(PA o k) mod fpt_length
  subset_i ← element }
foreach subset_i
  detect(subset_i)
return fpt[]

subroutine detect(subset_i)      // recover one bit from each subset
  total_count_0 = match_count_0 = total_count_1 = match_count_1 = 0
  foreach element in subset_i do
    if(Hash(0 o PA o k) mod γ equals 0) then {      // subset_0
      total_count_0 increment
      if(element is numeric)
        j = Hash(PA o k o 0) mod ε
        if(the j^th bit equals 0) then
           match_count_0 increment
      else if(element is textual)
        j = Hash(PA o k o 0) mod num_of_word_in_value
        if(Hash( the j^th word) is even) then
           match_count_0 increment }
    if(Hash(1 o PA o k) mod γ equals 0) then {      // subset_1
      total_count_1 increment
      if(element is numeric)
        j = Hash(PA o k o 1) mod ε
        if(the j^th bit equals 0)
           match_count_1 increment
      else if(element is textual)
        j = Hash(PA o k o 1) mod num_of_word_in_value
        if(Hash( the j^th word) is even)
           match_count_1 increment }
  if(match_count_0 > threshold(total_count_0, α)) or
  (match_count_0 < total_count_0 - threshold(total_count_0, α))   // pattern_0?
     return fpt[i] = 0
  else if(match_count_1 > threshold(total_count_1, α)) or
  (match_count_1 < total_count_1 - threshold(total_count_1, α))   // pattern_1?
     return fpt[i] = 1
  else return False            // no pattern
subroutine threshold(n, α)
  return minimum integer m that satisfies \sum_{k=m}^{n} C_n^k (1/2)^n < α/2
```

The selection process in our detection algorithm can be modeled as a Bernoulli trial, thus the match_count in a non-marked document is a random variable that meets a binominal distribution with parameters total_count and 1/2. Thus the threshold should satisfy (1) below.

$$P\{MATCH\_COUNT > threshold \mid total\_count\} + P\{MATCH\_COUNT < total\_count - threshold \mid total\_count\} < \alpha \quad (1)$$

Based on Agrawal's mathematic analysis [1], the threshold for a given total_count at confidence level of 1 - α can be calculated using formula (2) shown below..

$$\text{threshold} = \text{minimum integer m that satisfies} \sum_{k=m}^{total\_count} C_{total\_count}^{k} \left(\frac{1}{2}\right)^{total\_count} < \frac{\alpha}{2} \quad (2)$$

Thresholds for total_count from 1 to 30 when α = 0.01 are shown in figure 2 below. We can see that the bigger total_count is, the smaller portion of threshold is. Thus, given a large total_count, it gives the potential to resist attacks. For example, when total_count is 100, the threshold is only 64, which means with loss of nearly 40% of the mark, the fingerprint bit will still be detected successfully at the confidence level of (1 - α).

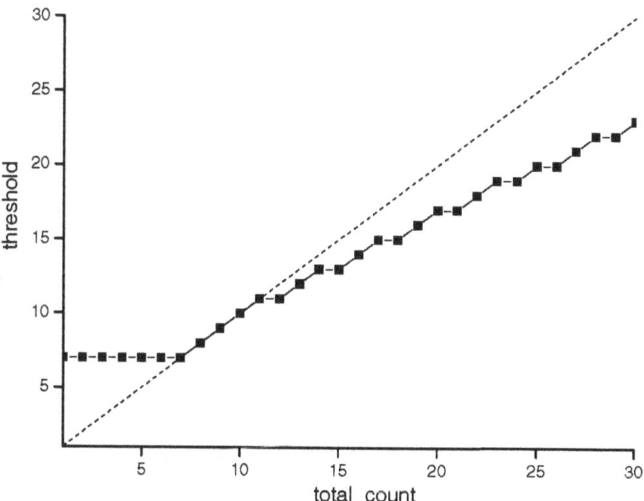


**Fig. 2.** The relationship between total_count and threshold when α = 0.01

## 3 Experiments and Analysis

We ran experiments in Windows 2003 with 2.0 GHz CPU and 512MB RAM. The XML data source is weblog.xml. For simplicity, we choose numeric elements to modify, results for textual elements are almost the same. We set = 10, = 3 and = 0.01, insert a 100-bit long fingerprint which can identify $2^{100}$ different distributors. We choose $N_1$ = 100,000 and $N_2$ = 10,000 of the records and experiment separately.

First we see our varied-size grouping method in figure 3, we list 10 groups. It shows that the total selected elements are almost *1/* of N and each group has varied but similar sizes. It means no element or few elements are selected in a certain group seldom happens. Some marks are expected in each group, thus we can have an entire fingerprint.

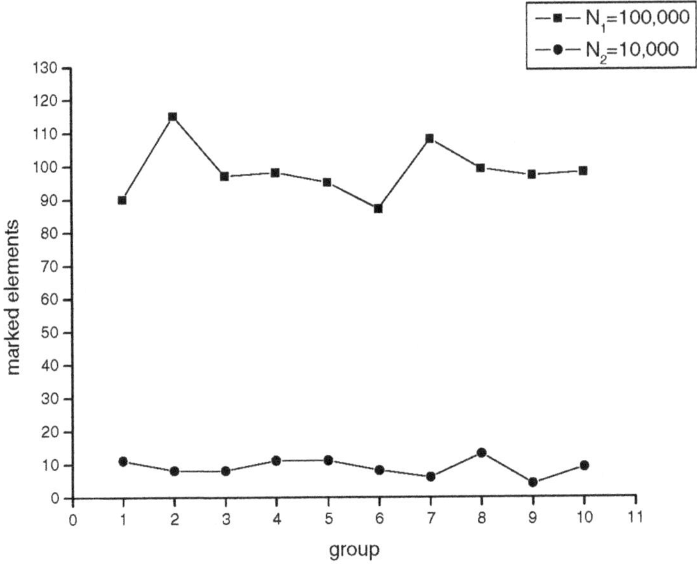


**Fig. 3.** Elements selected in each group

The situations in each group are alike with Wilfred Ng's selective approach [4]. So we can compare the amount of modifications in our scheme with Wilfred Ng's approach. We use the same secret key and the same other parameters to embed the same fingerprints. We can see in table 2 that the selected elements are all the same. Also the grouping results are the same. Because all parameters used in insertion are the same. When $N_1$ = 100,000, the elements needed to be modified are 4642 out of 9995 selected positions in our method. Compared with Wilfred Ng's method, it's 4960 elements to be modified out of 9995 selected positions. We reduce modifycations by 6.4%. When $N_2$ = 10,000, the elements needed to be modified are

**Table 2.** The amount of modifications compared with Wilfred Ng's selective approach

| | Modifications (our method) | Selected elements (our method) | Modifications (Ng's) | Selected elements (Ng's) | |
|---|---|---|---|---|---|
| $N_1$ = 100,000 | 4642 | 9995 | 4960 | 9995 | 93.6% |
| $N_2$ = 10,000 | 374 | 984 | 491 | 984 | 76.1% |

374 out of 984 selected positions in our method. Compared with Wilfred Ng's method, it's 491 elements to be modified out of 984 selected positions. We reduce modifications by 23.9%.

We can see a significant reduction of modifications when $N_2$ = 10,000, not too significant when $N_1$ = 100,000. The reason is that when $N_1$ = 100,000, each group has about 100 elements; when $N_2$ = 10,000, each group has about 10 elements. It's nearer to 50% to be modified when N is bigger. It's like throw a coin. For example, if you throw 10 times, the number of times when head is up is fluctuating around 5 heavily. If increased to 100 times, the number of times when head is up will be near 50. So the bigger N is, the reduction of modifications is less significant, see figure 4, we show 10 groups.

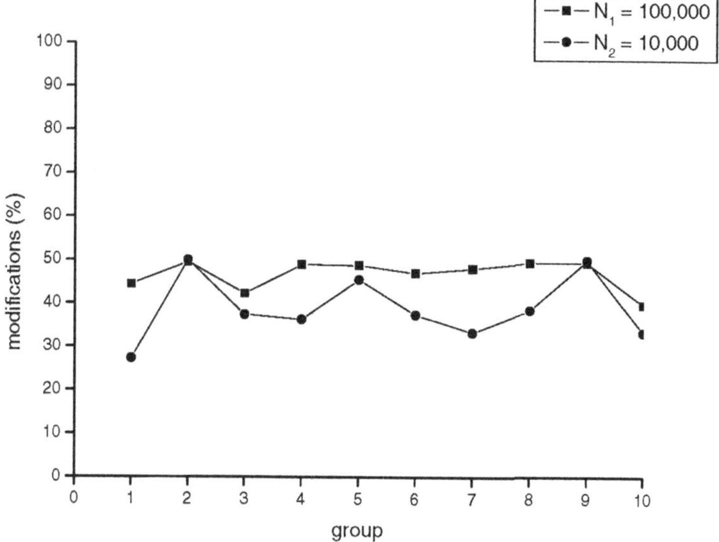


**Fig. 4.** Portion of elements needed to be modified

We can look at each group as an XML document input in Wilfred Ng's selective approach, thus we can compare the robustness level of our fingerprint bit in each group with Ng's approach. Because the confidence level is decided by the selected positions, so the robustness level of our fingerprint bit in each group is the same with Ng's result. We can see in table 2 that although the modifications are reduced by 23.9% when $N_2$ = 10,000, the number of selected positions are both 984 elements, so the robustness level maintains the same.

## 4 Summary

In this paper, we present our new watermarking scheme to embed fingerprints in XML data. Thus, we can not only prove ownership, but also identify illegal distributors since a unique fingerprint is embedded in each copy delivered to different distributors. We use a varied-size grouping method to preserve the order of the fingerprint's bits. To solve the problem of some categorical elements in XML document can't tolerant much modification, we make our effort to reduce modifications at the same insertion level, i.e., without bringing down the robustness of the fingerprint. In our scheme, to minimum modifications, we use the fingerprint to decide the inserting positions and then choose an inserting pattern. The experiments show that our scheme is effective to make less distortion to the original data and the fingerprint maintains the same robustness level at the same time. In some cases, we can reduce the modifications by 1/4.

In the future, we would like to research on the confidence level of the whole fingerprint, especially when part of the fingerprint has been destroyed; and how to argue ownership and identify illegal distributors from a fragmentary fingerprint.

## References

1. Rakesh Agrawal, Peter J. Haas, Jerry Kiernan.: Watermarking Relational Data: Framework, Algorithms and Analysis. VLDB Journal. (2003).
2. Radu Sion, Mikhail Atallah, Sunil Prabhakar.: Rights Protection for Relational Data. Proceedings of ACM SIGMOD. (2003) 98–109.
3. David Gross-Amblard.: Query-preserving Watermarking of Relational Databases and XML Documents, PODS 2003, San Diego CA. (2003)191–201.
4. Wilfred Ng and Ho-Lam Lau.: Effective Approaches for Watermarking XML Data. DASFAA 2005, LNCS 3453, pp. 68–80, 2005.
5. Yingjiu Li, Vipin Swarup, Sushil Jajodia.: Constructing a Virtual Primary Key for Fingerprinting Relational Data. DRM'03, Washington, DC, USA. 2003.
6. Radu Sion, Mikhail Atallah, Sunil Prabhakar.: Resilient Information Hiding for Abstract Semi-Structures. Proceedings of IWDW. 2004.
7. Radu Sion, Proving Ownership Over Categorical Data. Proceedings of ICDE 2004, 2004.
8. Yingjiu Li, Huiping Guo, Sushil Jajodia.: Tamper Detection and Localization for Categorical Data Using Fragile Watermarks. DRM'04, Washington, DC, USA. 2004.

# A Novel Labeling Scheme for Secure Broadcasting of XML Data

Min-Jeong Kim, Hye-Kyeong Ko, and SangKeun Lee*

Department of Computer Science and Engineering,
Korea University, Seoul, South Korea
{cara2847, ellefgt, yalphy}@korea.ac.kr

**Abstract.** With the fast development of the Web, a web document source periodically broadcasts its document to multiple users. The web document could contain the sensitive information and it should be sent to users who have an authority accessing the sensitive information. In a well-known method, called *XML Pool Encryption*, the sensitive information is separated from the document, and then, it is encrypted. Therefore, reconstruction of a document is required when the document is shown to a user. For the reconstruction, it is very important that we identify the location of decrypted information effectively and efficiently. In this paper, we propose a labeling scheme to support the fast reconstruction of document, based on the use of encryption techniques. The proposed labeling scheme supports the inference of structure information in any portion of the document. In the experimental results, our labeling scheme shows an efficiency in searching for the location of decrypted information.

## 1 Introduction

In the Web environment, XML (eXtensible Markup Language) [13] is rapidly becoming the standard for data representation and exchange. XML data to be broadcasted via the Web could contain between *public* information (all users can see) and *sensitive* information (users who have the authority can see). On this account, the demands for secure broadcasting of XML data are increasing. Secure broadcasting of data means that only a user who has the authority for the sensitive information can see that. In order to secure XML data, researches related to XML security have been studied [3], [5], [11], [12]. In particular, nodes (i.e., XML elements) that contain the sensitive information are selected, and then the nodes are moved into a pool and encrypted in the *XML Pool Encryption* approach [5]. A pool of encrypted nodes and unselected nodes are broadcasted to multiple users. Each user could decrypt according to authority of oneself. Because encrypted nodes were separated from the original XML document, the reconstruction of document should be performed.

* Corresponding author.

R. Nayak and M.J. Zaki (Eds.): KDXD 2006, LNCS 3915, pp. 95–104, 2006.

### 1.1 Motivation

The biggest problem of document reconstruction is to search for locations of decrypted nodes. We should know location of nodes in the document to solve this problem. In order to know location, the labels of nodes are needed. We can obtain the relation of nodes through the label of nodes. To label an XML document, the *XML Pool Encryption* [5] adopts the Modified Adjacency List Mode (MALM) labeling scheme. In the MALM, the range of the ancestor node label include the range of the descendant node label. By using the range-based labeling scheme, we can easily identify the ancestor-descendant relationship among nodes. However, searching for the exact location of each node is difficult, because it only represents the range of node label. In this paper, we propose a new labeling scheme that labels a child node by extending the parent's label to represent the structural information of XML document. The proposed labeling scheme provides an easy identification of relationships among nodes (ancestor-descendant as well as parent-child relationship).

The rest of the paper is organized as follows. Section 2 presents related work of this paper. Section 3 defines the proposed labeling scheme. Section 4 details the performance study and analysis by comparing the proposed labeling scheme with the MALM. Section 5 presents conclusion and future work.

## 2 Related Work

The problem with security for XML document is increasingly gaining attention [3], [4], [11]. In relation to XML security, the World Wide Web Consortium (W3C) is working on XML security standards. It provides a set of technical standard to meet security requirements. XML Signature Working Group created a specification for defining digital signatures [12] in an XML format. In addition, the XML Encryption Working Group of W3C [11] is developing a process for encrypting/decrypting XML documents and XML syntax is used to represent the encrypted content. In the W3C XML Encryption [11], if the contents are overlapped, the same portions of XML document could be re-encrypted for multiple users (Super-Encryption). However, it is not possible to encrypt an ancestor node while leaving any of the descendants of this node. Also, in Super-Encryption, there might be an information leakage between different users regarding their capabilities when compared each other [5].

Christian Geuer-Pollmann [5] proposed the idea of bringing the property from XML access control, which provides the granular access to an XML document [2], [3] [4], to the XML Pool Encryption. It focuses on how to encrypt an XML document at the granularity of nodes. The idea of *XML Pool Encryption* is to encrypt each node separately and to move all encrypted nodes from their original location in the document into a pool of encrypted nodes. The granularity of encryption is provided which is different from the subtree-based encryption of W3C XML Encryption.

# 3 A Novel Labeling Scheme to Handle Secure Broadcasting of XML Data

In this section, first, we show an example about the identification of relationships among nodes and present the architecture of secure broadcasting for XML data. Then, we propose a new labeling scheme to identify the relationship among nodes. The proposed labeling scheme supports easy handling of node relationships.

**Example 1.** Company C provides various contents to their subscribers. They broadcast two kinds of contents: paid contents and free contents. The subscribers who pay the money can see paid contents. The payment records that who paid the subscription rates are managed by the company. Fig. 1 shows "Contents" example. The "game" contents in sports, and the "story" and "picture" contents in culture are paid contents. User A who pays for the "sports" contents could see both free contents and sport contents. User B pays for the "sports" and "art" contents. The paid contents are moved into the pool and encrypted. The free contents and pool are broadcasted to multiple subscribers.

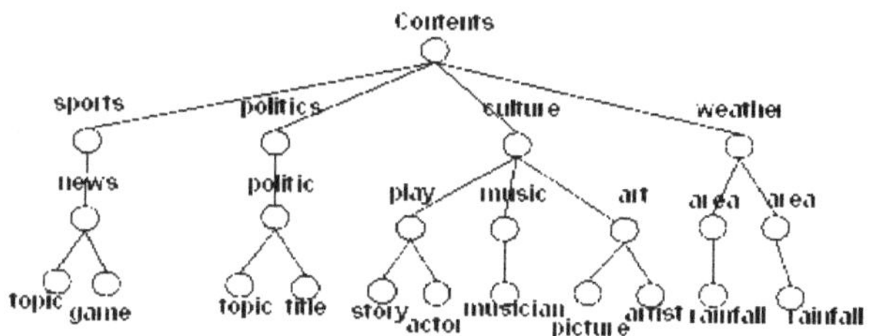


**Fig. 1.** The example of XML data

## 3.1 An Architecture of a Secure Broadcasting System

In this section, we present a system architecture with the above example. We select the sensitive nodes (paid contents) according to Payment Record and these are moved into a *pool*. Then, the nodes are encrypted separately by own keys.

**Payment Record.** We record about who charged the money for the paid contents, because the user to pay the money has the access authority about paid contents. The authorities of users to see the sensitive nodes are represented using XPath [10]. For example, authorities of User A and B are represented like that: *Contents/sports/news/game*, *Contents/culture/art/picture*.

**Key Management.** We encrypt the sensitive nodes in order to avoid the access of users who do not have the authority accessing to the sensitive nodes. The keys are assigned to each node.

The document (free contents and encrypted paid contents) is broadcasted to multiple users. We assume that the keys to decrypt the paid contents and Payment Record are transferred to user devices through a secured channel.

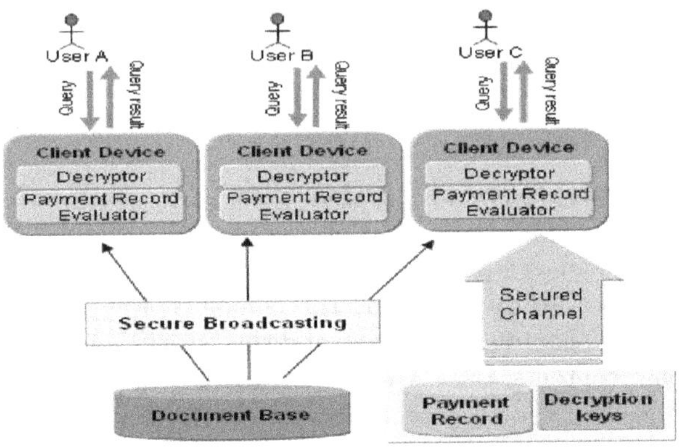


**Fig. 2.** The whole process of secure broadcasting system

Fig. 2 illustrates an architecture of a secure broadcasting system, where payment record evaluator evaluates the record, and then the document is decrypted using relevant keys. When requested, reconstruction of the document is desired to view the portion with the access authority.

### 3.2 The Proposed Labeling Scheme

The proposed labeling scheme is capable of identifying the relationship among nodes easily.

**Labeling Construction.** The labels of all nodes are constructed by three significant components *(C1, C2 and C3)*, which are unique.

1. *Level component(C1)* - It represents the level of node in the XML document. The level of the tree from root to leaf is marked such that the level of root is 1.
2. *Inherited label component(C2)* - The component that succeeds to the label of parent node, which eliminates the level component from a parent node label, is inherited. In succeeding the label of the parent node, the exact location of the node can be identified.
3. *Sibling order component(C3)* - It represents the relative location among the sibling nodes. A unique label is created by three components, which are concatenated by a “delimiter (.)”.

**Labeling Scheme for XML Document.** The labeling for an XML document is divided into root node and internal nodes.

**Definition 1.** (Label for root node $r$) *The root node is the first level. Because it does not have a sibling node and parent node, the value of the second component is null. In addition, values of the first component and the third component are 1, respectively.*
$L(x) = C1_{root\ r} \cdot C2_{root\ r} \cdot C3_{root\ r} = 1.nil.1$

**Definition 2.** (Label for internal node $x$) *C1 is represented by the level of corresponding node. C2 is created by inheriting the parent node label which eliminates the level component. Lastly, C3 represents the order of sibling nodes.*
$L(x) = C1_{current\_node\ x} \cdot C2_{current\_node\ x} \cdot C3_{current\_node\ x}$

1. $C1_{current\_node\ x}$ = *level of current node* $x$
2. $C2_{current\_node\ x}$ = *concatenate* $C2_{parent\_node}$ *and* $C3_{parent\_node}$
3. $C3_{current\_node\ x}$ = *sibling order of* $C3_{current\_node}$

The below Fig. 3 is the labeled XML tree by applying the above definitions.

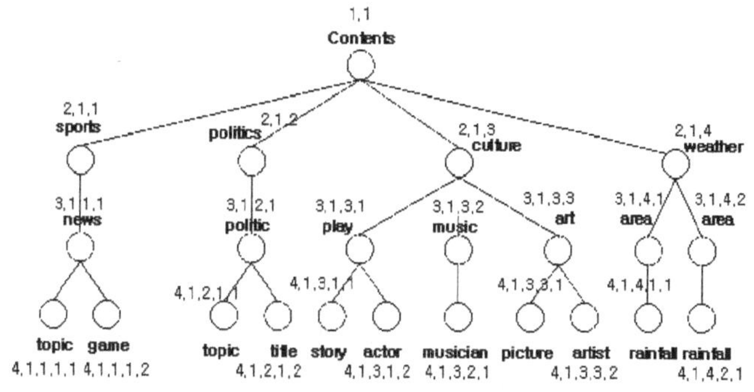


**Fig. 3.** Labeled XML data

**The Location of Node in XML Tree.** Identifying the relationship among nodes is essential to searching for the proper location of encrypted nodes in the *pool.* We propose a node labeling scheme to quickly identify the relationship among nodes.

**Lemma 1.** (Parent-child relationship ) *If node $x$ is a parent node of node $y$, the two labels satisfy the following.*

1. $C1_{parent\_node\ x}$ = *decrease in the level of the* $C1_{child\_node\ y}$
2. $C2_{parent\_node\ x}$ = *substring of* $C2_{child\_node\ x}$ *which eliminates the final part of* $C2_{child\_node\ x}$
3. $C3_{parent\_node\ x}$ = *the final part of* $C2_{child\_node\ x}$
4. $C2_{child\_node\ y}$ *is a string which is concatenated into* $C2_{parent\_node\ x}$ *and* $C3_{parent\_node\ x}$

**Lemma 2.** (Ancestor-descendent relationship) *If node $x$ is an ancestor node of node $y$, the two labels satisfy the following.*

1. $C1_{ancestor\_node\ x} < C1_{descendant\_node\ y}$
2. $C2_{ancestor\_node\ x}$ = *substring of* $C2_{descendant\_node\ y}$ *which corresponds to the length of* $C1_{descendant\_node\ y} - C1_{ancestor\_node\ x}$
3. $C3_{ancestor\_node\ x}$ = *the last substring of a part of* $C2_{descendant\_node\ y}$ *which corresponds to* $C1_{ancestor\_node\ x}$
4. $C2_{ancestor\_node\ x} \subset C2_{descendant\_node\ y}$

**Lemma 3.** (Sibling relationship) *If node $x$ and node $y$ are sibling nodes, the two labels satisfy the following.*

1. $C1_{right\_node\ x} = C1_{left\_node\ y}$
2. $C2_{right\_node\ x} = C2_{left\_node\ y}$
3. $C3_{right\_node\ x} = C3_{left\_node\ y} + 1$

**Algorithm 1** Searching a location of node

```
Input: L(p_n) - label of nodes p_1,p_2, p_n in the pool,
       L(x) - label of comparing node x
Output: relationship between node p_n and node x
For(i<n){
  Cr is the comparing value of C1_node_pi and C1_node_x
  if(Cr == 1) // parent-child relationship
    compare the C2_node_pi and L(x) that is eliminated C1 from L(x);
     if (the values are identical)
      node x is a parent of node p ;
  else if(Cr == 0) // sibling relationship
    if(C1 and C2 of node x and y, respectively, are identical &&
    C3_node_x is bigger than C3_node_p by one)
      node x is a preceding sibling node of node p;
  else if(Cr > 1) // ancestor-descendent relationship
    if(C2_node_x and C3_node_x is included in C2_node_p)
      node x is an ancestor node of node p;
  else //no relationship
    there is no relationship among nodes;
}
```

### 3.3 Labeling Scheme for Secure Broadcasting of XML Document

In this section, first, the process of *XML Pool Encryption* is focused on, then, the proposed labeling scheme is applied to *XML Pool Encryption* in order to search for the location of nodes. Algorithm 1 describes how to find the proper location of nodes in the *pool*. The relationship of nodes is identified, using the above algorithm when a user A requests to view an XML document. After identifying relationship among nodes, the XML document is reconstructed.

## 4 Performance Evaluation

We have implemented the proposed labeling scheme and performed an experiment in order to estimate the performance. We evaluated the time taken to search for exact location of decrypted nodes and compared it with the MALM labeling scheme used in the *XML Pool Encryption.*

### 4.1 Experimental Environment

According to the Algorithm 1, we implemented the labeling scheme using Java 2 [9] and XML Security Suite [6]. We used XMark [8] and DBLP [7] dataset to generate XML documents. For XMark dataset, we selected various scaling factors (0.001 ~ 0.06) to create from 0.1 MB to 6.9 MB of documents. For DBLP dataset, we used IBM Generator [1] to create from 0.025 MB to 1.4 MB of documents. For selecting the nodes to be encrypted, we used XPath [10]. Table 1 shows the XPath expression used to represent nodes to be encrypted.

**Table 1.** The Locations of node to be encrypted

| **XMark** | |
|---|---|
| P1 | /africa/item/* |
| P2 | /africa//*/text//* |
| P3 | //*/text/keyword |
| P4 | item/mailbox/mail/date |
| P5 | parlist/listitem |
| P6 | parlist/listitem/text |
| P7 | parlist/listitem/text/keyword |
| **DBLP** | |
| P1 | title//*/sub |
| P2 | sub/sup |
| P3 | sub/sup/tt |
| P4 | sub/sup/tt/ref |

### 4.2 Evaluation Results

To evaluate the performance of the proposed labeling scheme, we measured the location searching time of the nodes in the pool. In order to observe the relationship between the location searching time and the number of nodes, we measured the number of nodes to be compared while identifying the location of encrypted nodes.

**The Number of Nodes to Be Encrypted.** Fig. 4 shows the number of nodes to be encrypted according to the size of XML documents. In Fig. 4(a) and 4(b), the number of encrypted nodes is maximized at P5 and P1 for XMark and DBLP dataset, respectively. The number of encrypted nodes increases when the size of XML document increases.

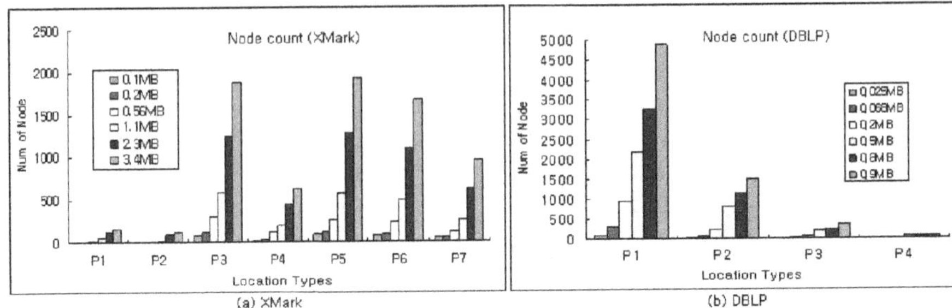


**Fig. 4.** The number of nodes to be encrypted

**Table 2.** The number of nodes to be compared

| | XMark (0.1MB) | | | DBLP (0.025MB) | |
|---|---|---|---|---|---|
| | Proposed scheme | MALM | | Proposed scheme | MALM |
| P1 | 12 | 84 | P1 | 94 | 6,197 |
| P2 | 3 | 21 | P2 | 31 | 2,064 |
| P3 | 79 | 22,452 | P3 | 9 | 701 |
| P4 | 20 | 2,514 | P4 | 3 | 61 |
| P5 | 91 | 42,572 | | | |
| P6 | 80 | 36,306 | | | |
| P7 | 55 | 24,557 | | | |

**The Number of Nodes to Be Compared to Search the Location in a XML Document.** To compare the location searching time of the proposed labeling scheme with the MALM, we observed the number of encrypted nodes to search for a location in XML document and measured the location searching time. The number of nodes to be compared, when searching for the location, is shown in Table 2. According to various location types, expressed by XPath, the number of encrypted nodes and the number of nodes to be compared are different. The above results demonstrated that the number of nodes to be encrypted is related to the number of compared nodes to search for the location, and the location searching time is affected. In the proposed labeling scheme, the label is not compared with another label of nodes because the label of the parent and ancestor nodes can be inferred from node, which searches for the location.

The result presented in the Fig. 5(a) and 5(b) indicate that the performance of the proposed labeling scheme outperforms the existing MALM, on all size of XML documents for location type P1 and P3 of XMark dataset and P1 and P4 of DBLP dataset. This is because, with the MALM approach, the number of encrypted nodes increases rapidly in all XML document sizes, therefore the number of comparison of the label among nodes increases.

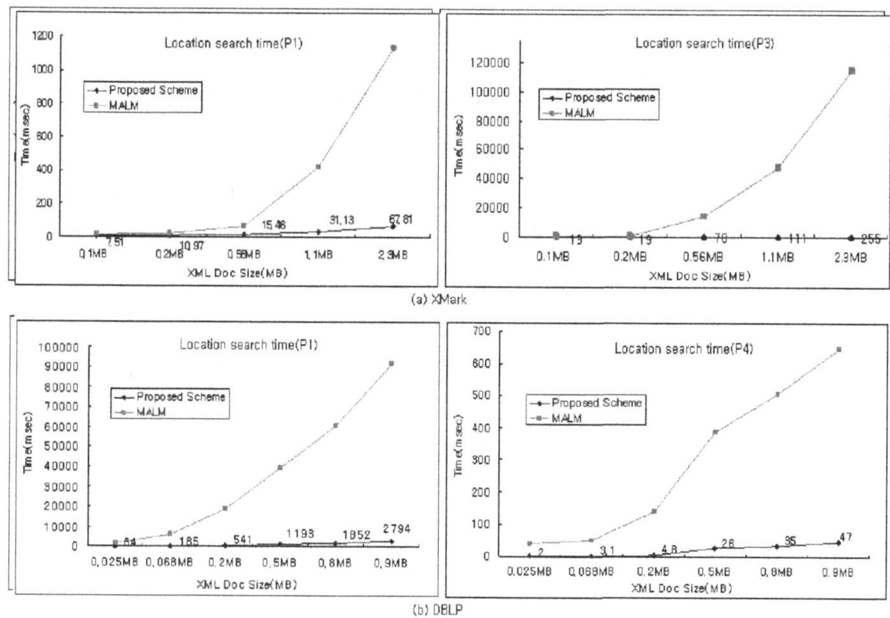


**Fig. 5.** Labeling time comparison according to the XML document size

## 5 Conclusion

With the advent of XML as a standard for data representation and exchange over the Internet, the issues for security of XML are of paramount importance. As the demands for security mechanisms are increased, W3C launched the XML Encryption [11] working group in 2001, and proposed a specification for XML Encryption. However, the W3C XML Encryption does not allow for encrypting an ancestor, while leaving a descendant, because it supports the subtree based encryption. In order to support the granularity of encryption, a pool encryption is proposed in the *XML Pool Encryption* [5], encrypting each node separately and moving all encrypted nodes from the XML document into a pool of encrypted nodes.

In this paper, we proposed a novel labeling scheme for secure broadcasting of XML document over the Internet. Similarly to the *XML Pool Encryption*, encrypted nodes in the *pool* are broadcast with the unencrypted nodes of the XML document to multiple users. In order to search for the proper location of encrypted nodes in an XML document, the effective and efficient identification of relationship among nodes has been presented. In particular, the proposed labeling scheme supports easy handling of the "parent-child", "ancestor-descendant", and "sibling" relationships of nodes. In the proposed scheme, the labels of nodes in an XML document contain the information regarding their parent and ancestor nodes as succeeding of the portions of the parent node label. Therefore, comparison with all labels of other nodes is not required when identifying relationship among nodes. The results of the experimental study are presented to evaluate the performance of the proposed labeling scheme and the MALM. The proposed labeling scheme is superior to the MALM in terms of the number of nodes used to search for the proper location of a node, and the location searching time according to the XML document.

We plan to devise other representation of the proposed labeling to reduce the overhead, e.g., by converting string type into integer type.

## References

1. S. Abiteboul, P. Bunneman, and D. Suciu. *Data on the Web:From Relations to Semi-structured Data and XML.* Morgan Kaufmann, 1999.
2. E. Bertino, S. Castano, E. Ferrari, and M. Mesiti. Controlled access and dissemination of xml document. In *ACM Web Information and Data Management*, pages 22–27, 1999.
3. E. Damiani, S. D. C. di Vidercati, S. Paraboschi, and P. Samarati. Securing xml documents. In *EDBT*, pages 121–135, 2000.
4. E. Damiani, S. D. C. di Vimercati, S. Paraboschi, and P. Samarati. A fine-grained access control system for xml documents. *ACM Transactions on Information and System Security*, 5(2):169–202, 2002.
5. C. Geuer-Pollmann. Xml pool encryption. In *ACM Workshop on XML Security*, pages 1–9, 2002.
6. IBM. Xml security suite. http://www.alphawowks.ibm.com/tech/xmlsecuritysuite.
7. M. Ley. Dblp database web site, 2000. http://informatik.uni-trier.de/ley/db.
8. A. Schmidt, F. Wass, M. L. Kersten, M. J. Carey, I. Manolescu, and R. Busse. Xmark : A benchmark for xml data management. In *VLDB*, pages 974–985, 2002.
9. Sun. Java, 2005. http://java.sun.com.
10. W3C. Xpath. http://www.w3c.org/TR/XPath.
11. W3C. Xml encryption wg, 2001. http://www.w3.org/Encryption/2001/.
12. W3C. Xml-signature syntax and processing, 2002. http://www.w3.org/TR/xmldsig-core/.
13. W3C. extensible markup language (xml) 1.0, 2004. http://www.w3.org/TR/REC-xml/.

# Author Index

# Lecture Notes in Computer Science

For information about Vols. 1–3817

please contact your bookseller or Springer

Vol. 3925: A. Valmari (Ed.), Model Checking Software. X, 307 pages. 2006.

Vol. 3924: P. Sestoft (Ed.), Programming Languages and Systems. XII, 343 pages. 2006.

Vol. 3923: A. Mycroft, A. Zeller (Eds.), Compiler Construction. XV, 277 pages. 2006.

Vol. 3922: L. Baresi, R. Heckel (Eds.), Fundamental Approaches to Software Engineering. XIII, 427 pages. 2006.

Vol. 3921: L. Aceto, A. Ingólfsdóttir (Eds.), Foundations of Software Science and Computational Structures. XV, 447 pages. 2006.

Vol. 3920: H. Hermanns, J. Palsberg (Eds.), Tools and Algorithms for the Construction and Analysis of Systems. XIV, 506 pages. 2006.

Vol. 3916: J. Li, Q. Yang, A.-H. Tan (Eds.), Data Mining for Biomedical Applications. VIII, 155 pages. 2006. (Sublibrary LNBI).

Vol. 3915: R. Nayak, M.J. Zaki (Eds.), Knowledge Discovery from XML Documents. VIII, 105 pages. 2006.

Vol. 3905: P. Collet, M. Tomassini, M. Ebner, S. Gustafson, A. Ekárt (Eds.), Genetic Programming. XI, 361 pages. 2006.

Vol. 3904: M. Baldoni, U. Endriss, A. Omicini, P. Torroni (Eds.), Declarative Agent Languages and Technologies III. XII, 245 pages. 2006. (Sublibrary LNAI).

Vol. 3903: K. Chen, R. Deng, X. Lai, J. Zhou (Eds.), Information Security Practice and Experience. XIV, 392 pages. 2006.

Vol. 3901: P.M. Hill (Ed.), Logic Based Program Synthesis and Transformation. X, 179 pages. 2006.

Vol. 3899: S. Frintrop, VOCUS: A Visual Attention System for Object Detection and Goal-Directed Search. XIV, 216 pages. 2006. (Sublibrary LNAI).

Vol. 3896: Y. Ioannidis, M.H. Scholl, J.W. Schmidt, F. Matthes, M. Hatzopoulos, K. Boehm, A. Kemper, T. Grust, C. Boehm (Eds.), Advances in Database Technology - EDBT 2006. XIV, 1208 pages. 2006.

Vol. 3895: O. Goldreich, A.L. Rosenberg, A.L. Selman (Eds.), Theoretical Computer Science. XII, 399 pages. 2006.

Vol. 3894: W. Grass, B. Sick, K. Waldschmidt (Eds.), Architecture of Computing Systems - ARCS 2006. XII, 496 pages. 2006.

Vol. 3890: S.G. Thompson, R. Ghanea-Hercock (Eds.), Defence Applications of Multi-Agent Systems. XII, 141 pages. 2006. (Sublibrary LNAI).

Vol. 3889: J. Rosca, D. Erdogmus, J.C. Príncipe, S. Haykin (Eds.), Independent Component Analysis and Blind Signal Separation. XXI, 980 pages. 2006.

Vol. 3888: D. Draheim, G. Weber (Eds.), Trends in Enterprise Application Architecture. IX, 145 pages. 2006.

Vol. 3887: J.R. Correa, A. Hevia, M. Kiwi (Eds.), LATIN 2006: Theoretical Informatics. XVI, 814 pages. 2006.

Vol. 3886: E.G. Bremer, J. Hakenberg, E.-H.(S.) Han, D. Berrar, W. Dubitzky (Eds.), Knowledge Discovery in Life Science Literature. XIV, 147 pages. 2006. (Sublibrary LNBI).

Vol. 3885: V. Torra, Y. Narukawa, A. Valls, J. Domingo-Ferrer (Eds.), Modeling Decisions for Artificial Intelligence. XII, 374 pages. 2006. (Sublibrary LNAI).

Vol. 3884: B. Durand, W. Thomas (Eds.), STACS 2006. XIV, 714 pages. 2006.

Vol. 3881: S. Gibet, N. Courty, J.-F. Kamp (Eds.), Gesture in Human-Computer Interaction and Simulation. XIII, 344 pages. 2006. (Sublibrary LNAI).

Vol. 3880: A. Rashid, M. Aksit (Eds.), Transactions on Aspect-Oriented Software Development I. IX, 335 pages. 2006.

Vol. 3879: T. Erlebach, G. Persinao (Eds.), Approximation and Online Algorithms. X, 349 pages. 2006.

Vol. 3878: A. Gelbukh (Ed.), Computational Linguistics and Intelligent Text Processing. XVII, 589 pages. 2006.

Vol. 3877: M. Detyniecki, J.M. Jose, A. Nürnberger, C. J. '. van Rijsbergen (Eds.), Adaptive Multimedia Retrieval: User, Context, and Feedback. XI, 279 pages. 2006.

Vol. 3876: S. Halevi, T. Rabin (Eds.), Theory of Cryptography. XI, 617 pages. 2006.

Vol. 3875: S. Ur, E. Bin, Y. Wolfsthal (Eds.), Haifa Verification Conference. X, 265 pages. 2006.

Vol. 3874: R. Missaoui, J. Schmidt (Eds.), Formal Concept Analysis. X, 309 pages. 2006. (Sublibrary LNAI).

Vol. 3873: L. Maicher, J. Park (Eds.), Charting the Topic Maps Research and Applications Landscape. VIII, 281 pages. 2006. (Sublibrary LNAI).

Vol. 3872: H. Bunke, A. L. Spitz (Eds.), Document Analysis Systems VII. XIII, 630 pages. 2006.

Vol. 3870: S. Spaccapietra, P. Atzeni, W.W. Chu, T. Catarci, K.P. Sycara (Eds.), Journal on Data Semantics V. XIII, 237 pages. 2006.

Vol. 3869: S. Renals, S. Bengio (Eds.), Machine Learning for Multimodal Interaction. XIII, 490 pages. 2006.

Vol. 3868: K. Römer, H. Karl, F. Mattern (Eds.), Wireless Sensor Networks. XI, 342 pages. 2006.

Vol. 3866: T. Dimitrakos, F. Martinelli, P.Y.A. Ryan, S. Schneider (Eds.), Formal Aspects in Security and Trust. X, 259 pages. 2006.

Vol. 3865: W. Shen, K.-M. Chao, Z. Lin, J.-P.A. Barthès (Eds.), Computer Supported Cooperative Work in Design II. XII, 359 pages. 2006.

Vol. 3863: M. Kohlhase (Ed.), Mathematical Knowledge Management. XI, 405 pages. 2006. (Sublibrary LNAI).

Vol. 3862: R.H. Bordini, M. Dastani, J. Dix, A.E.F. Seghrouchni (Eds.), Programming Multi-Agent Systems. XIV, 267 pages. 2006. (Sublibrary LNAI).

Vol. 3861: J. Dix, S.J. Hegner (Eds.), Foundations of Information and Knowledge Systems. X, 331 pages. 2006.

Vol. 3860: D. Pointcheval (Ed.), Topics in Cryptology – CT-RSA 2006. XI, 365 pages. 2006.

Vol. 3858: A. Valdes, D. Zamboni (Eds.), Recent Advances in Intrusion Detection. X, 351 pages. 2006.

Vol. 3857: M.P.C. Fossorier, H. Imai, S. Lin, A. Poli (Eds.), Applied Algebra, Algebraic Algorithms and Error-Correcting Codes. XI, 350 pages. 2006.

Vol. 3855: E. A. Emerson, K.S. Namjoshi (Eds.), Verification, Model Checking, and Abstract Interpretation. XI, 443 pages. 2005.

Vol. 3854: I. Stavrakakis, M. Smirnov (Eds.), Autonomic Communication. XIII, 303 pages. 2006.

Vol. 3853: A.J. Ijspeert, T. Masuzawa, S. Kusumoto (Eds.), Biologically Inspired Approaches to Advanced Information Technology. XIV, 388 pages. 2006.

Vol. 3852: P.J. Narayanan, S.K. Nayar, H.-Y. Shum (Eds.), Computer Vision – ACCV 2006, Part II. XXXI, 977 pages. 2006.

Vol. 3851: P.J. Narayanan, S.K. Nayar, H.-Y. Shum (Eds.), Computer Vision – ACCV 2006, Part I. XXXI, 973 pages. 2006.

Vol. 3850: R. Freund, G. Păun, G. Rozenberg, A. Salomaa (Eds.), Membrane Computing. IX, 371 pages. 2006.

Vol. 3849: I. Bloch, A. Petrosino, A.G.B. Tettamanzi (Eds.), Fuzzy Logic and Applications. XIV, 438 pages. 2006. (Sublibrary LNAI).

Vol. 3848: J.-F. Boulicaut, L. De Raedt, H. Mannila (Eds.), Constraint-Based Mining and Inductive Databases. X, 401 pages. 2006. (Sublibrary LNAI).

Vol. 3847: K.P. Jantke, A. Lunzer, N. Spyratos, Y. Tanaka (Eds.), Federation over the Web. X, 215 pages. 2006. (Sublibrary LNAI).

Vol. 3846: H. J. van den Herik, Y. Björnsson, N.S. Netanyahu (Eds.), Computers and Games. XIV, 333 pages. 2006.

Vol. 3845: J. Farré, I. Litovsky, S. Schmitz (Eds.), Implementation and Application of Automata. XIII, 360 pages. 2006.

Vol. 3844: J.-M. Bruel (Ed.), Satellite Events at the MoDELS 2005 Conference. XIII, 360 pages. 2006.

Vol. 3843: P. Healy, N.S. Nikolov (Eds.), Graph Drawing. XVII, 536 pages. 2006.

Vol. 3842: H.T. Shen, J. Li, M. Li, J. Ni, W. Wang (Eds.), Advanced Web and Network Technologies, and Applications. XXVII, 1057 pages. 2006.

Vol. 3841: X. Zhou, J. Li, H.T. Shen, M. Kitsuregawa, Y. Zhang (Eds.), Frontiers of WWW Research and Development - APWeb 2006. XXIV, 1223 pages. 2006.

Vol. 3840: M. Li, B. Boehm, L.J. Osterweil (Eds.), Unifying the Software Process Spectrum. XVI, 522 pages. 2006.

Vol. 3839: J.-C. Filliâtre, C. Paulin-Mohring, B. Werner (Eds.), Types for Proofs and Programs. VIII, 275 pages. 2006.

Vol. 3838: A. Middeldorp, V. van Oostrom, F. van Raamsdonk, R. de Vrijer (Eds.), Processes, Terms and Cycles: Steps on the Road to Infinity. XVIII, 639 pages. 2005.

Vol. 3837: K. Cho, P. Jacquet (Eds.), Technologies for Advanced Heterogeneous Networks. IX, 307 pages. 2005.

Vol. 3836: J.-M. Pierson (Ed.), Data Management in Grids. X, 143 pages. 2006.

Vol. 3835: G. Sutcliffe, A. Voronkov (Eds.), Logic for Programming, Artificial Intelligence, and Reasoning. XIV, 744 pages. 2005. (Sublibrary LNAI).

Vol. 3834: D.G. Feitelson, E. Frachtenberg, L. Rudolph, U. Schwiegelshohn (Eds.), Job Scheduling Strategies for Parallel Processing. VIII, 283 pages. 2005.

Vol. 3833: K.-J. Li, C. Vangenot (Eds.), Web and Wireless Geographical Information Systems. XI, 309 pages. 2005.

Vol. 3832: D. Zhang, A.K. Jain (Eds.), Advances in Biometrics. XX, 796 pages. 2005.

Vol. 3831: J. Wiedermann, G. Tel, J. Pokorný, M. Bieliková, J. Štuller (Eds.), SOFSEM 2006: Theory and Practice of Computer Science. XV, 576 pages. 2006.

Vol. 3830: D. Weyns, H. V.D. Parunak, F. Michel (Eds.), Environments for Multi-Agent Systems II. VIII, 291 pages. 2006. (Sublibrary LNAI).

Vol. 3829: P. Pettersson, W. Yi (Eds.), Formal Modeling and Analysis of Timed Systems. IX, 305 pages. 2005.

Vol. 3828: X. Deng, Y. Ye (Eds.), Internet and Network Economics. XVII, 1106 pages. 2005.

Vol. 3827: X. Deng, D.-Z. Du (Eds.), Algorithms and Computation. XX, 1190 pages. 2005.

Vol. 3826: B. Benatallah, F. Casati, P. Traverso (Eds.), Service-Oriented Computing - ICSOC 2005. XVIII, 597 pages. 2005.

Vol. 3824: L.T. Yang, M. Amamiya, Z. Liu, M. Guo, F.J. Rammig (Eds.), Embedded and Ubiquitous Computing – EUC 2005. XXIII, 1204 pages. 2005.

Vol. 3823: T. Enokido, L. Yan, B. Xiao, D. Kim, Y. Dai, L.T. Yang (Eds.), Embedded and Ubiquitous Computing – EUC 2005 Workshops. XXXII, 1317 pages. 2005.

Vol. 3822: D. Feng, D. Lin, M. Yung (Eds.), Information Security and Cryptology. XII, 420 pages. 2005.

Vol. 3821: R. Ramanujam, S. Sen (Eds.), FSTTCS 2005: Foundations of Software Technology and Theoretical Computer Science. XIV, 566 pages. 2005.

Vol. 3820: L.T. Yang, X.-s. Zhou, W. Zhao, Z. Wu, Y. Zhu, M. Lin (Eds.), Embedded Software and Systems. XXVIII, 779 pages. 2005.

Vol. 3819: P. Van Hentenryck (Ed.), Practical Aspects of Declarative Languages. X, 231 pages. 2005.

Vol. 3818: S. Grumbach, L. Sui, V. Vianu (Eds.), Advances in Computer Science – ASIAN 2005. XIII, 294 pages. 2005.